Julius Hensel

Macrobiotic or our diseases and our remedies

For practical physicians and people of culture

Julius Hensel

Macrobiotic or our diseases and our remedies
For practical physicians and people of culture

ISBN/EAN: 9783742818782

Manufactured in Europe, USA, Canada, Australia, Japa

Cover: Foto ©ninafisch / pixelio.de

Manufactured and distributed by brebook publishing software
(www.brebook.com)

Julius Hensel

Macrobiotic or our diseases and our remedies

MACROBIOTIC

OR

OUR DISEASES

AND

OUR REMEDIES.

FOR PRACTICAL PHYSICIANS AND PEOPLE OF CULTURE

BY

JULIUS HENSEL
PHYSIOLOGICAL CHEMIST.

TRANSLATED BY PROF. LOUIS H. TAFEL
OF URBANA UNIVERSITY O.

Kreatin $C_4 N_3 H_9 O_2 H_2 O$ Sarkin $C_5 N_4 H_4 O H_2 O$ Xanthin $C_5 N_4 H_4 O_2 H_2 O$ Harns. Ammon. $C_5 N_4 H_4 O_3, N H_3$

Cum Deo!

FROM THE SECOND REVISED GERMAN EDITION.

————→‖←————

PUBLISHED BY BOERICKE & TAFEL
1011 ARCH STREET, PHILADELPHIA.

PREFACE.

As the introduction to an opera gathers together the motives which are the foundation of the musical work, so I wish herewith to prepare the way by stating that I have throughout ascribed the origin of internal diseases to a *diminished electric force.* This is in agreement with the demonstrated unity of forces, which I have also applied to our vital force. The cause of this diminution of the electric force may be found either in *respiration of oxygen insufficient* in itself, or in the *more difficult absorption* of the quantity of oxygen required for the prosecution of our vital functions, owing to a diminished number of red blood corpuscles, or in strong *emotions of the mind,* or in *atmospheric influences,* or in the reduction of the nervous tension in special regions of the body owing to a *partial check in the circulation of the blood.* As a tangible cause of the diminution of the blood disks which absorb the oxygen, I have pointed to the insufficiency of the amount of sulphur, lime and iron contained in many articles of food; but I have made especially prominent the loss of electrifying blood-salts which loss is not sufficiently compensated by food, as being by far the most frequent cause and one hitherto not sufficiently considered, producing a diminished electric tension of the nerves and thence fatal maladies. This daily loss of blood-salts is thoroughly natural, because the *urea* resulting as a product from respiration, *requires mineral salts in order to combine with them into permanent combinations of double salts.* If the urea does not find enough mineral salts in the blood, it is changed by the chemical absorption of water from the venous blood into carbonate of ammonia, which produces paralysis of the nerves, blood-poisoning and even leads to putrescence. Various lesser as well as more intense degrees of chronic sufferings are connected with this state. Either singly or together the afore-mentioned injurious influences manifest themselves in anaemia, chlorosis, and dyscrasic affections, in pulmonary consumption, dropsy, rheumatism, diabetes, cramps and in inflammatory conditions.

As to urea ($CO\ N_2\ H_4$), its materials are found in part in an unoxidized form as a double stratum of *gelatine sugar* (COO, CHH, $NHHH$), and in part in a grouping of the atoms into *cyanide of ammonia* within the molecules of the bases of flesh: *Creatin*, Sarkin, Xanthin and ureate of ammonia (s. p. 67) as their physiological foundation.

Thus may be understood, why we and our children are subject to the consequences which the liberation of Ammonia from the stagnating blood or from urea carries with it in the form of *catarrhal affections*, whenever our blood in its serum does not contain the natural protection and defense in a due quantity of mineral salts. What a luminous significance is there contained in this with respect to the children's catarrh, called "Diphtheria"! This begins with stagnation of the blood in the Thymus Gland, whence the products of the putefraction of the albumen of the blood are spread with their infecting force over the whole vascular system. Now as the children of a family are in the same state as to nutrition, only the first case of diphtheria is needed in order to carry away the whole troop of children and also the mother. For when the mother is confined to the sick-bed of her children, the regular progress of the household is checked. Salted milk-soups and flour-soups are not provided. But the respiration takes its usual course and in consequence urea is liberated, which carries off the blood-salts in the excretion of the urine. These salts deficient before, are diminished from day to day. One child dies; a second, badly cared for has to take to its bed; the mother watches over each, and in her grief she does not think of eating and drinking, and she follows her children and dies because no physician prescribes for her the physiological salt-water which would save her.

We cannot, however, deny, that a certain guarantee against certain affections is offered by a vegetarian diet, because it is richer in mineral elements than a meat-diet; especially if the nourishing vegetables are cultivated not in a garden fertilized with stable manures, but on natural mountain soil, or at least on fields which are preserved in their pristine fertility by manuring with finely ground rocks; but also vegetarians remain subject to the injuries caused by changes of weather, and by emotions, which affect the nervous system, causing stagnation of the blood and the pathologic consequences flowing from it. Besides this, vegetarians also are subject to hereditary tendencies, which may be restored to their normal state by improving the constitution of the blood by means of preparations of iron, lime and sulphur, as we see e. g. in scrofulosis; for when the blood is renewed the whole body can be built up anew.

While I have endeavored, to furnish an explanation of the chemical processes taking place in certain pathologic conditions, I hope thus also to have furnished to practising physicians a sort of thread of Ariadne, which may be of service to direct them in the labyrinth of diseases that has hitherto prevailed, and which may considerably enlarge the boundaries of the healing art.

Hermsdorf unterm Kynast.

THE AUTHOR.

PUBLISHER'S PREFACE.

The first edition of Hensel's Macrobiotic was published in Germany in the year 1882. At first its success seemed doubtful. It was, and is, an original work, a radical departure from the prevailing ideas and theories held by men of science, the book of an original investigator and thinker.

The new comer was received with ridicule by those whose minds were fixed in the old ideas and theories; others, more cautious but equally conservative, shook their heads, but there was a sufficient number of untrammelled men to finally exhaust the first German edition, and to call for a new one. The author placed the manuscript and coprights for the new, thoroughly revised and enlarged edition of the Macrobiotic, and also that of a new work, *Das Leben, seine Grundlage und die Mittel zu seiner Erhaltung,* ("Life, its Foundation and the Means for its Preservation") in our hands for us to bring out both works in German and to have them translated into English.

Both works are now completed and have been published in Germany; copies of the German editions may be obtained through the book-trade, and the first work, The Macrobiotic, is herewith placed before the English speaking public.

The reasonings, theories and deductions contained in the Macrobiotic are decidedly original and may work a revolution in many of the departments of Science and Medicine. That they are not mere idle ideas is evidenced by the fact that in Germany there are several establishments in operation, involving a considerable investment of capital, where profitable and practical demonstration is afforded that Hensel's theories have passed from the category of theory to that of applied science. In these establishments Hensel's theory, that the stones which have hitherto been regarded a mere encumbrance of the earth may be turned to great use and profit in scientific fertilizing, is practically demonstrated and there is steadily increasing demand for their product.

The new ideas have also made great progress in the domain of medicine. In this-field their applicability is almost limitless.

THE PUBLISHERS.

CONTENTS.

—

INTRODUCTION

When I published the first edition of this work ten years ago, I did so in order to supply medical practitioners with what was at that time altogether wanting, a clear view of the chemical processes which occur in the human body, and in doing so to render a service to mankind at large. I was encouraged to undertake the task by the consideration that I was in advance of the majority of medical practitioners, from having commenced the study of medicine in later life after having been an apothecary for twenty years, and thence equipped with chemical and physical knowledges. My position on commencing the study, therefore, differed from that of the young students who enter on it fresh from the gymnasium, and I was thence able to see many things in a clearer light than that in which they were represented by the professors. One thing which especially struck me was the view taken by teachers of medicine that the general facts of chemistry could not be applied in physiology. "The human body is not a retort." With this dictum the most important, fundamental, and universally accepted chemical facts seemed to me to be set aside, preference being given to unfounded hypotheses rather than to assured physical and chemical facts. As long as one gropes in the dark, it stands to reason that a thousand mistakes will be made, but as soon as light is shed around, the real bearing of things can be grasped. Chemistry furnishes us, in the case to which I refer, with just this light, enabling us to solve what are apparently the most difficult problems, by means of a few simple figures and equations.

That Chemistry could not be sooner utilised in the advancement of medical science is explained by the fact that the laying of the first foundation of physiological knowledge is but of very recent date. It is only one hundred and fifteen years ago, that the Englishman Priestley and the Frenchman Lavoisier discovered, in 1774, that we live by means of a chemical process of combustion, our bodily substance uniting with the air which we inhale, and yielding the products of combustion which we exhale as aqueous vapour, carbonic acid gas, and nitrogen, the chemical action exactly corresponding to that which we find in the case of a burning candle or a lamp fed with oil. If we cease to inhale we shall be suffocated, that is to say, our spark of life is extinguished, just as the flame of a petroleum lamp is extinguished if we prevent the air from passing to it. As the inhalation of combustible air and the exhalation of exhausted air succeed one another, new material is supplied to replace that which is consumed, and so the action goes on, just as the streams which flow into the sea never cease to run, being fed by numberless brooks and springs which receive back again from the sea the water with which they part, by means of the vapour which under the action of the sun rises in the atmosphere.

The springs and rivulets which carry to our body new strength are the delicate chyle and lymph vessels, through which passes the fresh material derived from the chyme as it proceeds through the alimentary canal.

It seemed then to me that if there should be wanting in the alimentary matter supplied, certain constituent parts essential to the due cohesion of our bodily substance and to enable it to perform its functions,—in other words, should the material which is consumed and used up not be adequately replaced—our bodily and our psychical powers will gradually suffer a diminution, which, continued for a long time, will on a slight impulse inevitably in time assume the form of serious sickness.

What kind of constituent parts were to be considered in this matter, I never for a moment doubted. They could not be such as were capable of being resolved into gases, for such are merely the four substances Carbon, Oxygen, Hydrogen and Nitrogen, which we find contained in sufficient quantities in all alimentary matter, brandy, which contains no Nitrogen, only excepted; but then Providence has not given us brandy for food.

What had to be considered were the earthy constituents with which the above named four substances are chemically united and which, remaining as ash-products in the blood, after the consumption of the bodily substance by breathing, are continually thrown off dissolved in serum as urine. The natural method of healing which we shall here discuss, is based upon this consideration. In it nothing really novel is advanced, for the method has long been practised under the name of Folk-Medicine, though it has not received professorial recognition. Medical science has as regards its advancement laboured under a great disadvantage. Engineers, electricians, and the members of other professions come from the ranks of the scholars in the *realschulen*,*) and the sciences to which they have devoted themselves have advanced with giant strides during the last century. Medicine has, however, haughtily refused to admit to its study any scholars but such as come from the gymnasia**) (I was myself educated in a *realschule* and studied medicine abroad), and the science has consequently made no progress. To-day medicine, as

---

*) Scientific Schools.   **) Classical Schools.

taught in the schools, is no further advanced than it was in the time of barber Jenner, who in the year 1770, before Oxygen was known and consequently when no profit could be derived from the application of Chemistry, studied surgery and pharmacy in London. Jenner, while engaged in compounding the apothecary's drug called theriaca, learned that it had been invented by king Mithridates VI who, being of a suspicious turn of mind and living in constant dread of being poisoned, sought to counteract any attempts which might be made against his life by systematically inuring himself to poison, and so compounded theriaca. Such a story when once heard is never forgotten. Jenner did not forget it. While he was earning his bread in Berkeley by bleeding, by applying leeches, and by cupping, a peasant woman told him another story. She said to him, "milking cows which have cow-pox on the udder is a good thing to prevent one from taking small-pox. My dairy wenches have an eruption after such milking, but they never have the small-pox." These two stories suggested to Jenner the remarkable medical system of inoculating with vaccine, and to-day all medical men follow in his footsteps. They, being as ignorant as barber Jenner of chemical science, are worthy followers of a worthy leader. Humanity is to be made healthy by inoculation with putrefying matter! Has not the time surely come when the sickly art of medicine should be "inoculated" with some new healthy blood?

What is wanting in the physiology of to-day is, as I have before hinted, an acquaintance with the constituents of the ashes of our bodily substance which give fixity to the albuminous constituents. We are taught by professors that the human body consists of the before mentioned four elementary substances, Carbon, Oxygen, Hydrogen, and Nitrogen, the chemical symbols of which are, respectively, C, O, H, and N. This, in more intelligible chemical language, signifies that our bodies consist of Carbonic Acid, COO, and Ammonia $HHHN$, which contain at the same time the elements of water, OHH, and of oleine, CHH.

We are further told that our bodies consist of Chlorine, Cl, Fluorine, F, Phosphorus, P, and Sulphur, S. These substances can all be converted into gases. Mention is also, indeed, made of the earthy component-parts of our bodily substance, of Kalium, Natrium, Calcium, Magnesium, Iron, Manganese, and Silicon, but we are told that no knowledge has so far been attained of the part performed by these earthy substances in our organism, and they are, accordingly, set aside in considering the theory of life and of subsistence. "Man requires each day so much *fat*, so much *albumen*, so much *Carbo-hydrate*, and so much *water*." Thus teachers of Physiology, having no firm foundation on the earth, build up a system "in the air", and thus it is that they seek to heal us by Febrifuges, Antipyretics, Quinine, Morphia, Cocaïne, and Sulphonal, which are devoid of the earthy substances which maintain life, and consist only of spiritual or gaseous matter; and in truth these are hostile, evil spirits. So we are lifted from the earth to be strangled "in the air", as Antæus was by Hercules!

In such cases the blinder the professors are, the more instruments do they bring with them to the sick-bed. They come armed with a stethoscope, to hear with, a plessimeter to tap with, a thermometer, a sphygmograph, a probe forceps, ear and eye mirrors, a laryngoscope, jaw and nose mirrors, catheter, bougies, syringes, a microscope, a galvanic apparatus, and a chest of re-agents. And what is the result? At last *earth* has to heal everything.

In contrast to such a course, folk-medicine does not wait till the final end, before seeking earth as a remedy, but at the very commencement resorts to earth and ashes, as, for instance, when curing fever or cramp.

In cases of intermittent fever people drink urine, which contains the ashes and the salt constituents which remain as the products of the combustion of our bodily substance in breathing, and which, when absorbed by the lymphatic channels, restore to the body its strength and tension. Real ashes, however, using the word in its common sense, are still used by the people as medicine, and with success.

The ashes of burnt magpies and those of the hoof of the elk are, it is said, used even now in the Diakonissen-Anstalt in Dresden, in the treatment of epilepsy, though this remedy has disappeared out of the pharmacopœia. Such ashes are, indeed, of service, for both bird's feathers and the hoofs of ruminants contain sulphur, and it is this which, with the lime of the bones and the alkali and soda combinations of the blood, soothingly helps to balance the action of the exciting, phosphoric, ashless nerve-matter. To obtain this curative material, it is true, we need not burn magpies or the hoofs of elks. The ashes of the leaf of the beech-tree may be used with just as much success, containing as the leaf, does all the ashy constituents required alike by man and the elk. Does not the elk eat beechleaves and build up its bodily substance there from?

There is this difference between mammals and worms—the first require earthy substances in larger quantities than do the latter.

That earthy substances are serviceable in epilepsy is shown by the fact that insufficiency of salts and earths leads to cramp. Napoleon the First, for instance, who had a restless brain, was subject to epileptic attacks. Both his restlessness and his epilepsy may be attributed to his diet, for he used to eat daily the brain of a newly slaughtered ox. Sulphur and lime are wholly wanting in the brain substance, but these constituents are absolutely necessary for the formation of the red blood-corpuscules with which the nerve quieting Oxygen chemically combines.

Napoleon also suffered from the itch. The cause of this was the same. Through a deficiency of Sulphur, part of the mucous membrane penetrated by the terminations of nerves became changed into worms (itch-mites), just as worms are produced in the intestines of sucklings when the mother's milk, or that supplied from the cow, is poor in lime and sulphur.

Lime and sulphur, indispensable to man, are injurious to worms. The silk-worm furnishes us with a proof of this. This worm lives, not upon beech-tree leaves,—which, being rich in earthy substances, would be poison to it,—but upon mulberry leaves which contain but little of such substances. How great is the difference in this respect between these two leaves may be seen from a comparison of the constituents of their ashes.

	Sulphuric Acid	Phosphoric Acid	Silicic Acid	Potash	Soda	Lime	Magnesia
Beech leaves .	21	24	195	30	3	258	34
Mulberry-leaves	1	12	41	23	0	3	6

These figures are significant, showing as they do that beech-leaves yield five times the amount of earthy substances contained in mulberry leaves; together with the pregnant fact that the lime contained in beech-leaves is 86 times as great, and the Sulphuric Acid 21 times as great as the quantity contained in mulberry-leaves. The more that lime and Sulphuric Acid are wanting, the more probable it is that worms will be produced.

Many facts appear mysterious when they are regarded by themselves, which can easily be explained when considered in relation with other facts. So we find many sucklings suffering at the same time from both worms and cramp, when they are supplied with diluted milk to which salt is not added. Milk contains only 6 to 7 thousandths of ash constituents whereas the human blood requires more than 8 thousandths of them. It will be obvious, therefore, that when, by admixture of equal parts of water, the salts contained in the milk are reduced to 3 thousandths, the lymph and blood constituted from this are unable to hold the nerve material within bounds. Worms consequently breed in the intestines. The deficiency of salt in the blood with respect to the nerves, for reasons which will be explained later on, causes cramp. I may here incidentally mention that a deficiency in the amount of the ash constituents contained in the food results in scrofula and rickets. By these examples I wish merely to point out how imperatively necessary it is that we should recognise the importance to our organism of the mineral tensional matter. As long as this is not recognised by the professors, so long must that painful uncertainty exist which, alas! we now so frequently observe both with regard to the diagnosis and the treatment of diseases. It is not meant to be urged here that mineral substances, alone and exclusively, provide a panacea for all diseases. Such an assumption would be absurd. All that is here asserted is, that they must never be lacking when a certain degree of physical power of resistance is desired, and that they must in every case be kept in view. (It is certain, nevertheless, that there are other factors which play an important part in regard to health and disease. Among these are beside the diet,—climate, change of weather, nature of occupation, and, above all, the emotions. The last undoubtedly occupy the first place amongst factors causing disease, and we must not evade the consideration of them, but we shall find that their action also, amounts to an electro-chemical process. (It thus becomes the more incumbent on medical practitioners to study the physico-chemical structure of our organism from other standpoints than those which have hitherto been customary in Anatomy, Physiology, Pathology, and Therapeutics.

I

ANATOMICAL PART,

OR

THE STRUCTURE OF OUR BODILY SUBSTANCE.

"For thousands of years Medicine has found
remedies, but not a single truth, not a single law
of life."

Joseph Hyrtl's Lehrbuch der Anatomie
des Menschen. 1878. P. 21, Line 30.

According to the method of teaching which has hitherto prevailed,
the human body has been regarded as consisting of bones, ligaments,
sinews, muscles, arteries, absorbents, nerves, organs of sense, and viscera.
The viscera again are divided into organs of digestion, organs of breathing,
and organs of generation.

The digestive apparatus includes the oral cavity, the palate, the
teeth, the tongue, the salivary glands, the fauces, the œsophagus, the
stomach, the small intestine, the liver, the gall-bladder, the pancreas,
the spleen, the large intestine, and the peritoneum.

The respiratory apparatus includes the larynx with the vocal
ligaments, the windpipe, the thyroid gland, the lungs, and the pleura.

When there are so many different organs to be considered with
their functions, considerable time is needed. The matter is not, however,
so complicated as it appears at the first moment, and viewing it from
a chemical stand-point, we are enabled to trace in the entire human
organism a gratifying and complete unity.

How, indeed, could we hope to render the human organism intelligible by separating it into various distinct pieces? A clock placed
before us in fagments of damaged wheels, cylinders, cogs and springs, is
no longer recognisable as a clock, but is merely an unintelligible heap of
ruins. In the same manner the dissection of the body into separate parts,
into bones, muscles, blood vessels, glands and bundles of nerves, is not
calculated to promote the explanation of the simple life problem; on
the contrary, a thousand artificial conundrums are placed before us
instead of one single question.

Proceeding from a wrong hypothesis a man soon arrives at false conclusions. We must, if we are to arrive at a right understanding of a situation, place ourselves at its central point and, keeping that ever in view, observe what ways lead from it and whither they lead.

Hitherto the method of teaching Anatomy has been such as might have suited the time of Galen, when Chemistry was unknown. At the present day, however, the method is as unsuitable as is the inoculation method of barber Jenner.

I must here remark that Galen in teaching the structure of our bodies commenced with the skeleton, or scaffolding of bones, a method which we need not wonder at, for Galen's father was an architect and so required a "scaffolding" for every structure. That we have not even now, eighteen hundred years after Galen, and after the electric light of Chemistry has illumined the horizon during one hundred and fifteen years, advanced any further than this method, might afford matter for wonder, were it not notorious how difficult it is to efface wrong views which have been held for centuries and to place the simple truth in their stead. As an example of this I need only allude to the Copernican planetary system, and to Galileo.

The skeleton to serve as the basis of our knowledge of the bodily organism! Is it credible! The skeleton, which represents only a parody, a caricature of the human form! Taking it as a basis, the whole method of presentation founded upon it must be a caricature.

No, we must no longer go to the charnel-house for enlightenment as to the constituents of our living bodies. We must, on the contrary, proceed from the mobile nerve-substance, that matter which lays alike the foundation and provides the form. That these two elements—the foundation and the form of our bodies—apparently opposed to each other—should when viewed from a chemical standpoint, be found in hitherto unsuspected agreement, can only serve to support the view that all our bodily parts have their origin in one common fundamental material.

———

Anatomists have placed it to the credit of Goethe that his keen eye discovered that the skull consists of three modified vertebræ, and that he even recognised the peculiar characters of the vertebræ in the facial bones. Goethe himself says on this subject:

"So it was with the conception that the skull consists of vertebræ. The three posterior I soon recognised, but it was only in the year 1791, when I picked up, from the sand of the old Jewish cemetery at Venice, a battered sheep's skull, that I at once observed that the facial bones must also be derived from vertebræ."

The celebrated Peter Frank in his treatise on inflammation of the spine, also remarks that the skull may be regarded as the first vertebra, and the other vertebræ as so many skulls.

Such a view will not be combated here, for it is true that the vertebræ and the skull are similar in this respect, that they serve as protecting shells for the nervous matter enclosed within them. The fact is, however, that both Goethe and Peter Frank considered only the shell and not the kernel.

With regard to this kernel, I also, setting out from a consideration of the skull, have arrived at an independent view. I once looked down from an elevated seat in the anatomical theatre at the heap of bones provided for the lecture. Among these I noticed what appeared to be an ostrich egg, and 1 silently asked myself for what purpose it could have been placed among the bones. When, however, at the termination of the lecture, I inspected the egg more closely I discovered that I had made a mistake. The object was a thick oval shaped skull of a negro reposing on its apophyses, and when viewed from above, it could be easily mistaken for an ostrich egg.

From that day I have regarded the skull as a kind of egg-shell for the brain contained in it, and whether my fellow creatures be tall or short, 1 look upon them all as perambulating eggs.

The external resemblance of the human brain to the yolk of an egg is, indeed, striking in the highest degree. The correspondence of the blood-vessels in both the larger and smaller yolks from the ovary of a fowl with those of the arteries in the hard membrane covering the brain of a fœtus and the *Processus falciformis* of the *Dura mater* is very striking. I invite those of my readers who possess Heitzmann's Descriptive Anatomy, with illustrations, Vienna, 1875, to turn to figure 517, and to place the boiled yolk from the ovary of a hen beside it. They will be astonished at the absolute similarity.

But there is, further, no other essential difference between the egg-yolk and the brain. Do not the convolutions of the brain resemble the folds of an impregnated *vitellus* as one egg resembles another? And does not the development of the egg fully correspond with that of the brain into the nervous system?

Even in its matured form, the brain resembles a much furrowed egg-yolk, but one which has sent out a stalk or stem, the spinal cord, from which issue numerous ramifications which in part grow together again, and again separate, as, for instance, the nerve plexuses of arm and thigh. From all the ramifications finer branches issue which finally terminate in muscular tissue, a prototype of the division of an electric current for the service of a thousand incandescent lamps. The new-born child is, further, chiefly a mass of brain with merely a weak appendage. The egg-shell is still open. That the appendage grows pretty quickly and is, apparently, the only movable part, need not deceive us. That a skull-like bony substance is afterwards formed everywhere, and that numerous little brains (ganglia) are developed, like relay-stations, only confirms the main fact that all the additions are merely the continuation, the offshoots, the radii of the substance of which the great brain-egg is composed. At first the crawling egg, which we call the child, knocks its skull often enough against the floor, but the stem which has issued from it (the spinal column) with the branches, (arms and thighs) gradually becomes so strong that it supports the brain-egg upright, and secures it from tumbles, so that it no longer "knocks its head against the wall."

For the rest it is clear that the bones are not the chief thing. That the soft brain appears first is obvious. And is not the bone formed quite slowly and later on by the deposition of fine little stars of calcium phosphate in the glutinous substance of the cartilage? How foolish of us to allow ourselves to be imposed upon by the long and broad skeleton! To understand our organism we must take the nervous-system as our basis, and we may then construct the system of descriptive anatomy, which is now of so complicated a nature, on decidedly simpler and more harmonious lines. Before, however, I proceed in the following pages to perform such a task, I think it will be well to look critically at an extreme illustration of the old methods. If the structure of the human body is to be made intelligible, it will be necessary to consider the form of the whole as dominant, and the separate parts which may be detached from it must not be regarded as forming the constituent parts, any more than one should regard a tree as produced from the splinters which have been chipped off it with an axe. This comparison is such a close one that it continually impresses itself on me, but it has not done so on others who do not content themselves with saying that in our body we have to distinguish as separate parts, the head, heart, lungs, liver, spleen, kidneys, intestines, &c., but who, with regard to the principle of subdivision, lose themselves in the wildest confusion. For the theory has obtained that our body, as also that of animals, as well as the vegetable organism, is built up of separate, so-called cells, each of which leads a separate, independent existence, though all, in a certain way, are under one general rule, as are the members of a colony of corals. As a matter of fact such cells may be discovered with the microscope in several body tissues, having apparently circumscribed limits, but this does not by any means hold good of the entire body-tissue and especially not of the blood-plasma from which nevertheless cells issue. If then the formless plasma is the fundamental matter from which the cells originate, as snow-crystals originate in vapour, the cells cannot lay claim to be considered primary; we must rather assign the chief part to the mother substance, to the general fundamental material from which they spring, and take it for our first point of departure in endeavouring to arrive at a right understanding of organic growth and of the entire physiological process. Although the doing so will necessitate our somewhat trenching on the domains of Chemistry and of Physiology which I propose to treat of separately, a slight survey, even at this stage, may be of service.

It is necessary that we should understand by a cell and its contents a certain sum of movable component parts which, according to circumstances, may be smaller or greater. Instead, therefore, of calling it a cell, since what we have to consider is not the merely circumscribed space or an empty vessel but especially its contents, I would be understood to mean a mass of atoms which, from a denser grouping of a certain portion of its elements on the periphery (in accordance with the fundamental law of electricity that it always collects on the periphery, where it acts by attraction and condensation), appears to have fixed limits in relation to a contiguous similar mass of aggregated material.

Many of these circumscribed atomic heaps when looked at under the microscope are found to be of a spherical shape, while others have discoid, cubical, polyhedric, conic, cylindrical, hooped or fibre-shaped forms, according to the part of the body in which they are found. This difference of form is at once an indication that they are in a dependent relation to their surroundings and to the powers exerting influence on them.

As to the power which holds such atomic masses together for a certain time, not only in themselves but also with regard to each other, we have the universal law of attraction between oppositely electrified bodies and of repulsion between like electrified bodies.

In so far as the cell presents a shape, it has, of course, different sides, and consequently different opposite limits or poles. From this consideration we may easily deduce the cell's subordination to the ruling law, the result of such subordination being that the dissimilar poles come together, in other words, that there arise, in a certain direction, connected rows of similarly formed cells which may act as a central axis around which may be afterwards formed groups of differently shaped cells in layers, the cells resembling each other in every separate layer, and the layers covering the axis like a mantle, or, rather, like a galvanic casing, in the sense that such a casing must be regarded as composed of similar but not of exactly identical material.

We must here remark that this accumulation cannot go on ad infinitum either in the direction of length or on the periphery, but that the power of attracting further material finds its limit at the place where the zone of activity of the electric exciting centre ceases. It is of importance, in order to understand the nature of the electricity which accumulates on the periphery of a sphere or at the ends of a wire, that we recognise some central power of causation which propels the electric fluid centrifugally, for of course no action can originate in itself. Such a central point of departure governing the electric operations within the circle of its activity, as the sun governs the course

of the planets which gravitate around it, may be likened to the mathe-
matical centre of gravity of any given coherent mass. As this centre
of gravity even in the case of a hollow sphere is placed in its centre,
where we find none of the surrounding spherical matter, so I assume
the existence of an electric centre at the very point where no traces
of electricity are discernible.

The governing force of this electric centre acts for some distance
as a repelling power, but it is necessary, in order to make it possible
for its electric commands to be transmitted instantaneously to the next
atoms, that concentrated material should be at its disposal. Thus it
is that the radial repelling power of the electric centre is rendered
complete by means of its natural opposite, the power of attraction in
regard to bodily substance. A stick of sealing-wax rendered electric
by friction and so made capable of attracting little pieces of paper, is
an illustration of this.

Whereas the repelling power extends to a considerable distance,
the power of attraction is limited to the immediate proximity and to
a limited number of atoms which, like obedient servants of the mistress
of a house, perform what they are commanded. I think that we may
in this manner satisfactorily account for the diminutiveness of those
atomic masses called cells. The combining power of an electric
centre, with regard to loose individual atoms, only extends over a
certain zone, beyond which the government of another independent
electric centre begins. There is no reason for doubting that space
consists of such zones of connected electric activity, in other words,
there is no reason, why we should not assume that all mundane space
is occupied by such neutral centres of electricity governing certain
atomic masses which are in harmonious relation to other contiguous
atomic groups.

In order to better explain this and to apply the argument *ad ho-
minem*, let us consider the heart as a centre, not only on account of
its power of attraction but also on account of its repelling activity with
regard to the blood. The heart is not, indeed, a completely accurate
analogy, for it is by no means the centre of the organism, but is to be
rather regarded as the pole complemental of the brain, only coming
into existence by the growth of the brain mass and being formed out
of material attracted by it. The heart, although the later formed and
opposite complement of the originally sole existing brain material, must,
none the less, be regarded as finally dominant and decisive, for life
immediately ceases in the case of a stab in the heart, whereas such
cessation is by no means always the result of a stab or of a shot-wound
in the brain.

The heart may be said to be formed by the condensation, expansion,
and accumulation of substance from an arterial tube, the first delicate
arteries being formed by the coming together of bi-concave blood-

corpuscules in cylindrical arrangement, so that the electricity which is manifested at the periphery of the cylinders, in consequence of the compact accumulated material, exercises its power of attraction only on the smaller interior mass whilst incorporating its firmer constituent parts in the produced parietal space—in other words, a hollow cylinder is formed in which the watery substance, separated from the firm material, remains behind in the shape of a serous fluid.

If now, returning to the cells, which we have characterized as atomic masses, we examine them under the microscope we find that the circumscribed spaces are by no means hermetically sealed, on the contrary, that there is a lively intercourse between the interior of them and the outer world. Each of them by means of its own peculiar electricity, derived from the dissimilarity of the atoms massed within it, exercises power over a certain space, attracting other movable material coming within its reach and absorbing it, separating and repelling the useless material, and grouping in a suitable manner the applicable component parts, the result being that the atomic mass augments in bulk and density in its restricted space. The final result is (in accordance with my general law of concentration and resolution demonstrated by me in another work*) that the atomic mass, having become too dense, divides into two homogeneous parts, or into layers of a different nature.

We may explain the case of two new cells being formed from one by the fact that the compressed atoms in the first cell are no longer able to find sufficient space for free molecular movement. As the result of such a condition of things the similar electric poles and not the dissimilar electric poles of two molecules will in some places come together. Their mutual repulsion follows, and this action is extended giving its direction to the whole of the matter of the atomic mass, or the contents of the cell. The cell thus divides itself into two equal halves since the neutral matter accumulating more densely in the middle part, a partition-wall is formed, so that from one result two similar cells which, with newly constituted poles attached to each other, become joined to the general mass, so that an extension or lengthening of the axis by the insertion of a new link of the chain is thus rendered possible.

It may, however, also happen that several molecules in the interior of a cell may simultaneously give the impetus to re-grouping, so that inside the old cell 4, 6, 8, or even more independent groups, may arise which are separated from one another by neutral substance, symmetrically draw to themselves the remaining contents of the cell, and consume the material of the old cell wall. As in the former case we may find an explanation of the extension of the axis, so in this latter we may, to a certain extent, find the explanation of growth in the periphery.

---

*) Life: Its foundations and the means of its preservation &c. Published by Boericke & Tafel, Philadelphia-Leipzig.

In both cases the impulse to formation and renovation appears as an inborn law of life extending even to the smallest atomic masses.

An atomic mass or cell which is not thus productive (being prevented, perhaps, from being so by the want of such material as it can assimilate) loses its vital force. It may be regarded as moribund, comparatively dead, but it may, together with its torpid contents, be drawn into a process of rejuvenation and new life, by means of the action of a living and productive atomic mass into whose zone of electric activity it may have come.

From this presentation it will be observed that a certain collection of mobile atomic masses has an original endowment of impulse to productiveness. This law applies universally, even to stone dust, water, and air, from which plants are formed by the stimulating power of the electric sunlight.

In regard to the bodies of animals and especially the human body, the cohesion of its atoms mainly depends, on the one hand, on the electrically acting brain and its ramifications, and, on the other hand, on the action of the heart with its tubular radiations and its liquid contents. The human body lives only so long as the brain and the heart maintain their electric action.

The building up of our body proceeds, therefore, in accordance with the general law of growth and life, which is founded on impulse to formation (production) and to renovation or change of material, that is, the continual attraction and assimilation of a certain amount of new matter with a simultaneous or alternate rejection of a certain portion of the material already used or old.

An organism producing nothing, non-rejuvenating, not renewing itself by redintegration, and which has also no animating influence on its surroundings, becomes torpid and dies.

We are here setting forth from an independent stand-point, the theory of the human body, and we should ourselves pronounce judgment on our theory if we ourselves did not strive to animate, to electrify, and to assimilate. I would here like to ask, if impulse to production obtains in every single cell of our body, must not such an impulse manifest itself in the individual as a whole, in the impulse to create provided, and so long as, the individual is possessed of a sufficient amount of electricity?

Life, indeed, means *being electric, being productive, acquiring, forming, building up,* and *calling into existence.* Of such an electric state there are, of course, innumerable degrees, the natural result with regard to beings who are least electric, and so the most indolent, being that they succumb to fate, are treated as part of a dead mass, and, in spite of a passive resistance, are subjugated, led, ruled, and rendered tributary to others who possess more vital energy. This is exacted by the natural, fundamental law of life.

Having thus shown that the processes of life are dependent on mutual electric intercourse with the external world, it follows that the life process of each single individual may be affected in the most various manner by an external *vis major*, that is to say, the measure of the electricity originally given us may be augmented or diminished by external means. The whole store of electricity with which an individual is endowed may, in the fluctuation incident to growth and use, be gradually and in the course of a considerable time, imperceptibly expended ´ or the store may be squandered through the intense action of external forces in a short time, or, indeed, it may be exhausted in a single momentary explosion.

It will therefore be incumbent upon us, in the pathological portion of our inquiry, to consider, if we attach any importance to health and to the enjoyment of as vigorous and as long an existence as possible, by what means the store of electricity within the human organism may be preserved and augmented, and in what manner the store suffers the danger of being diminished and destroyed.

For the present we shall proceed to consider the structure of the human body, taking the structure as a whole, and not following the ordinary method of dividing the material respecting which instruction is given, and treating of it in pathological, medico-chirurgical, topographical, general, and descriptive anatomical sections—a system of treatment which has instead of promoting a clear insight, only resulted in confusion.

Morbid Anatomy should teach us the variations and changes to which the organs are liable in a state of disease; medico-chirurgic Anatomy, the relations of the organs to one another, and thus it should enable us to arrive at a knowledge of facts from the successive symptoms observed in cases of illness. Topographical Anatomy should safely guide the knife in the hand of the surgeon. General Anatomy treats of the delicate structure of the tissues of which the organism is composed; and, finally descriptive Anatomy treats of each organ separately, of its relation to other organs, gives to each its name, and tells us of the arrangement of the tissues which compose it. For the study of this last branch of anatomy, it is, we are told, necessary to dissect with certain instruments, so we are supplied with scalpel, forceps, scissors, hammer and saw, in order that we may isolate the different parts of the body from one another.

We find then that the anatomical teaching of to-day proceeds upon the same lines as did the teaching of three hundred years ago, when no one had the necessary knowledge of the chemical processes incident to our organism.

The machinery of our body must certainly be understood if we are anxious to preserve it in good condition, but it seems to me that such knowledge cannot be so well attained by dividing the subject into four or five different parts for study in different academic terms as by treating it as one comprehensive unity by viewing it from the chemical stand-point. By such a course alone is it possible to render the basis of the healing art a common possession, by diffusing a knowledge of the origin of diseases and of the methods of their prevention; rendering everyone, in the most important matters, independent and free from fear and prejudice; showing him what dangers and what erroneous paths are to be shunned; and leading him in the way to participate in the highest degree of human felicity.

THE EGG.

In the bird's egg which may be regarded as also typical of the basis of the mammalian organism and which on account of its proportionate size is studied most easily, we distinguish the "white" from the "yolk" which it encloses and surrounds. The former corresponds to the albumen of the blood, the latter to that of the nervous system. The white of the egg is soluble in water owing to its freedom from fatty constituents; the yolk on the other hand owing to the amount of fat which it contains and to the minute membranes which surround each of the groups of fatty particles, is only soluble in water in so far as it is intermixed with watery salt-solutions which, however, make up rather more than half its weight. In the white of the egg the content of water amounts to as much as seven eighths of the whole.

Both the white and the yolk contain potash, soda, lime, magnesia and iron; but while in the yolk these bases occur exclusively in combination with phosphoric acid, and silica, in the white they are besides this combined with sulphuric acid, hydrochloric acid and carbonic acid (or with sugar—a carbonate of the carbohydrate).

The yolk contains no chlorides or sulphates. Since however the yellow yolks are secreted in the ovary of the hen from the albumen of the blood, we conclude from the separation of the phosphates out of a complex of carbonates, sulphates, chlorides and phosphates as exemplified in the white and yellow of the egg, that:—

a) the phosphatic compounds stand evidently in adverse relationship to the carbonates, chlorides, and sulphates;

b) this very contrariety determines their correlation, since the opposing elements attract one another:

c) these *earthy substances* give evidence of their being *living forces* acting on one another both by the fact of their separating from one another, and of their remaining in adverse position on the opposite side of the membrane which encloses the yolk;

d) the silica which is common to both the compartments—the *white* and the *yolk* of the egg—acts as an electrically neutral mediating substance keeping up the state of electrical tension between the different parts by isolating them, and without its presence in certain amounts only incomplete relations could subsist between the white and the yolk; all these earthy constituents which together with the fatty and albuminous substances in their train keep up a living tension between one another I call the mineral *tension elements* of our bodily substance;

e) nearly one half of the egg yolk consists of fat *ergo* of combustible substances; in contradistinction thereto the white contains more than four fifths of its weight of water; this means that the combustible substance of the yolk is protected ·by a covering rich in water enclosing its membranes against the oxidizing tendency of the air; while to make security doubly sure, every separate fat particle of the yolk is covered by a protecting membrane—the result being, that any combination of the fat of the yolk with the outside air must take place in an extremely difficult manner through the membranes.

The further chemical characteristics of the yolk-fat and the white are closely related.

One of the most characteristic constituents of the phosphoric yolk-fat is a compound of saccharine matter with stearate of glycerine and phosphate of ammonia. The chemical union of the two latter substances is so intimate, that one could be equally justified in saying that the compound is an ammonium stearate combined with phosphate of glycerine and saccharine matter. In calculating, however, the chemical composition of the constituents of nerve fat (Lecitin=$C_{42} H_{84} O_9$ NP) it appears that two proportions of fat (or stearate of glycerine) are combined with phosphate of ammonia, thus;

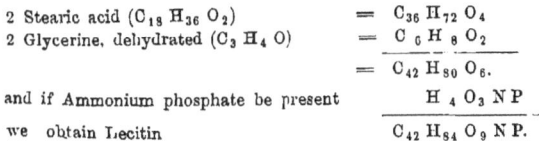

$$
\begin{array}{lll}
\text{2 Stearic acid } (C_{18} H_{36} O_2) & = & C_{36} H_{72} O_4 \\
\text{2 Glycerine, dehydrated } (C_3 H_4 O) & = & C_6 H_8 O_2 \\
\cline{3-3}
& = & C_{42} H_{80} O_6. \\
\text{and if Ammonium phosphate be present} & & H_4 O_3 NP \\
\cline{3-3}
\text{we obtain Lecitin} & & C_{42} H_{84} O_9 NP.
\end{array}
$$

The compound of Lecitin and sugar is called Protagon.

It is by the saccharine matter that the yolk-fat which corresponds to the brain fat (cerebrine) is related to the white of the egg. In fact it may be considered as a product of secretion from the latter, since the albumen of the animal organism ($C_{144} H_{112} O_{44} N_{13}$) consists essentially of equal proportions of grape-sugar (glucose) and "gelatine-sugar" (glycin) *plus* certain earthy and sulphur compounds, in accordance with the following calculation:

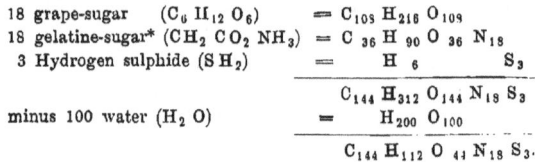

$$
\begin{array}{llll}
\text{18 grape-sugar} & (C_6 H_{12} O_6) & = & C_{108} H_{216} O_{108} \\
\text{18 gelatine-sugar* } & (CH_2 CO_2 NH_3) & = & C_{36} H_{90} O_{36} N_{18} \\
\text{3 Hydrogen sulphide } & (S H_2) & = & H_6 \qquad\qquad S_3 \\
\end{array}
$$

$$C_{144} H_{312} O_{144} N_{18} S_3$$
$$\text{minus 100 water } (H_2 O) \quad = \quad H_{200} O_{100}$$
$$C_{144} H_{112} O_{44} N_{18} S_3.$$

Animal albumen is nothing but aggregated vegetable albumen, which latter is formed in a very simple manner from partially oxidized saccharine matter which has combined with ammonia. I have in fact succeeded in preparing vegetable albumen by warming together 1 equivalent of tannin with 3 equivalents of grape-sugar until chemical combination took place. To the solution of the product in water I added 4 equivalents of ammonia together with 1 equivalent of sulphate of soda, and from the solution albumen could be precipitated in the well known manner.

Calculation shows that 1 equivalent of tannic acid ($C_{14} H_{10} O_9$) and 3 equivalents of grape-sugar ($C_6 H_{12} O_6$) with 2 equivalents of ammonia ($N_2 H_6$) after the detachment of water owing to chemical condensation, correspond to 4 equivalents of grape-sugar and 4 equivalents of gelatine-sugar (COO, CHH, NHHH) i. e. equal equivalents of both, and this relationship is likewise preserved in the case of animal albumen, which latter is consequently really and essentially accumulated vegetable albumen. The origin of the latter from partially dehydrated, and partially oxidized saccharine matter (tannin) together with ammonia, determines its close relationship to fat which is also produced from oxidized saccharine matter.

This connection between the white of egg or albumen and fat, which results from their both having their origin in the same material, is of the greatest importance both as regards our judgment as to the unity of our bodily substance and the conditions which govern its renewal; I am consequently of opinion that it would at this point be advisable to give the theoretic explanation of the various steps by which saccharine matter may be converted into albumen—under certain conditions into nerve-fat.

In the first place admit that sugar ($C_6 H_{12} O_6$) as pointed out in my work "Das Leben" has its various molecules arranged as shown in the figure at the side.

Grape-sugar

Such a particle of sugar on losing its water points, as an equivalent for its chemical condensation with like particles, or with other substances, takes the shape of a hexagon.

---

* Glycocoll or glycocine.

The chemical condensation with particles of the same kind thereupon takes place in such a manner that a group c_O^O from one residual molecule of sugar, attaches itself to a group $_H^H c$ of the neighbouring residual sugar molecule. This is determined without the necessity of any other explanation by the chemical affinity of the electro-positive hydrogen $_H^H$ for the electro-negative oxygen $_O^O$, and the chemical condensation in this way becomes comprehensible without requiring the presence of any material which should act as cement.

These remarks explain how it is that six hexagonal residual sugar molecules are able to group themselves around a seventh residual molecule, giving rise to the group drawn below consisting of 7 times C_6 H_8 O_4:

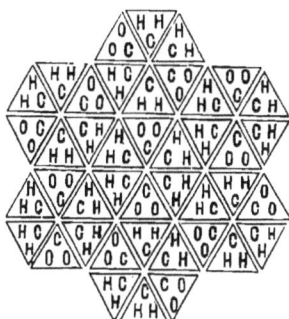

$$7 (C_6 H_8 O_4).$$

Since further in this group 12 atoms of oxygen are adjoined to the hydrocarbon molecules which belong to the central residual sugar molecule, and since all the 4 hydrocarbons giving off 4 molecules of water oxidize into of 4 molecules of carbonic acid c_O^O, it follows that 6 molecules of carbonic acid are now arranged round the centre including the 2 original molecules. But to balance this twelvefold oxidation 12 chemical decompositions or detachments of molecules must take place.

These 12 decompositions take place as follows. It is towards the circumference of the whole group that the electric discharge naturally takes place, and there 9 molecules of water are detached. In addition 3 other cleavages occur along lines running from the geometric centre symmetrically towards the periphery.

The result is the production of 3 molecules of tannic acid equal to 3 ($C_{14} H_{10} O_9$).

Tannic acid occurs in numerous parts of plants. Roots, bark, leaves and fruit contain it in huge quantities. In astringent fruits, however, it may again disappear by taking up water from the flowing sap, and combining therewith under the influence of the sun's heat to re-form saccharine matter and carbonic acid. The carbonic acid then inflates the ripening fruits and makes them capable of floating on water as may be observed in the case of the better sorts of apples.

In addition my view of the arrangement in space of the atoms in the tannic acid is confirmed by the circumstance that when heated in carbonic acid it decomposes into chinon (C_6 H_4 O_2) and pyrogallol (C_6 H_6 O_3)

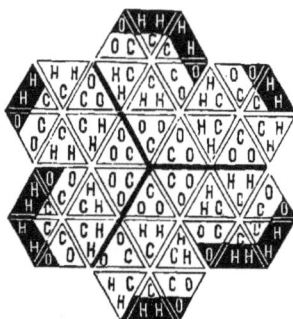

3 Tannic acid=(C_{14} H_{10} O_9)

The original group of 7 residual sugar molecules can however experience a vastly different fate, when instead of existing in the leaves or bark where evaporation takes place, it occurs dissolved in the rising cambial sap. In this case not tannic acid but olein is produced as follows:

The presence of water and the exclusion of atmospheric air has for a result that the oxidation of the 7fold molecule of saccharine matter can only take place as a consequence of reactions going on inside of it. Such are the loss of oxygen from carbonic acid which results from its change into oxalic acid, or that, according to which 6 molecules of carbonic acid combine with 18 of water, and separating 12 hydrogen peroxide are converted into saccharine matter. In such cases besides tannin, lecitin may be simultaneously produced provided ammonium phosphate be present. For this is required $4 \times 12 = 6 \times 8$ hydrogen peroxide, which decomposes into 6×8 water and 6×8 oxygen.

The addition of only 8 atoms of oxygen to the 4 central hydrocarbon molecules of the 7fold sugar molecule converts them however not into carbonic acid but into formic acid. C_{HO}^{HO}.

This is a circumstance of decisive importance; for in such a case where the oxidation of the central hexagon gives rise to formic acid, hydrogen peroxide is produced at each of the 6 edges, 4 times by oxidation and twice through combination of C_O^O with $_H^H C$ of a residual sugar molecule at the periphery.

All that is now required is for a ray of the sun to enter through the transparent membrane in order to effect a complete chemical separation, as an equivalent for the operation of light.

It is at the same time comprehensible that nothing can be detached at the edges except what is present there i. e. hydrogen peroxide.

But since the effect of the sun's rays is by no means confined to the central hexagon, but like lightning passing along a lightning conductor, strikes upon *all* the edges also those of the *peripheral* hexagons, it will be plain that for *every* one of the edges (i. e. 7×6 edges) an equivalent amount of force is demanded, and must be paid for in the same coin as at the commencement namely in hydrogen peroxide.

In the latter way originates either simply olein or under special conditions the nerve-fatty substance—lecitin.

Ordinary olein or plant stearoptin is formed according to the following calculation in which it is assumed that sufficient water is present to supply each of the 18 peripheral edges with 3 molecules of water:

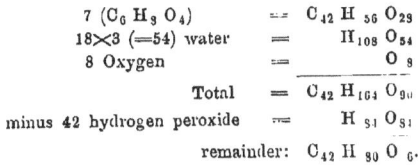

$$
\begin{aligned}
7 \ (C_6\,H_3\,O_4) \quad &= \quad C_{42}\,H_{56}\,O_{29} \\
18 \times 3 \ (=54) \text{ water} \quad &= \quad H_{108}\,O_{54} \\
8 \text{ Oxygen} \quad &= \quad O_8 \\
\hline
\text{Total} \quad &= \quad C_{42}\,H_{164}\,O_{90} \\
\text{minus 42 hydrogen peroxide} \quad &= \quad H_{84}\,O_{84} \\
\hline
\text{remainder:} \quad &\quad C_{42}\,H_{80}\,O_6.
\end{aligned}
$$

In this formula we recognize without much difficulty a double molecule of stearate of glycerine-anhydrite: $2 \ (C_{18}\,H_{36}\,O_2 + C_3\,H_4\,O)$.

From the latter results lecitin through combination with ammonium phosphate as already explained.

Two points must here be taken into consideration. Firstly the hydrogen peroxide thus set free is instantly decomposed into water and oxygen, so that the freshly produced lecitin has command immediatly on its formation of a supply of oxygen serving to vivify it, and consequently the theory of tannin and lecitin at the same time includes that of the original generation of animal life—in so far at least as we take the articulated animals into consideration.

Since we have in leaves containing chlorophyl both sugar and albumen existing side by side, a double reaction between the phosphate of lime and the ammoniacal albumen is all that is required to give rise to ammonium phosphate. Then, however, no insect's puncture is necessary, but the detaching of the hydrogen peroxide is adequate to explain the production of living lecitin whenever an atom of phosphate of lime is carried along in the rising sap. Thus doubtless originated the larva of the gallfly. And so also at the present day the phosphatic flour, when oxidized in a moist state and becoming putrid, produces the various kinds of mites. Furthermore spontaneous generation of wood-lice takes.

place in the sugar-containing buds, of phylloxera in root tendrils, of the larvae of the bark-scarabs in the sweet cambial sap, and of the caterpillar germs in the tannic-acid-containing sap of the leaves.

I regard lecitin or nerve-fat as the essential and indispensible basis of the more intelligent animal life. The greater the purity in which this basis occurs the higher the intelligence of the animal even in the smallest space. The superior acuteness of the senses displayed by bees, ants, flies &c, is due to this fact. The more however that one or another of the essential elements from which lecitin is compounded is absent, and the greater the extent to which a rôle is played by other bodies, the more clumsy, soulless, and stupid does the creature become.

Pure lecitin requires for its production as already pointed out: 7 particles of sugar ($C_6 H_{12} O_6$), 1 particle of ammonium phosphate 8 of oxygen, 42 of water, and sunshine.

If sunshine be absent only mites emerge from the moist flour.

If but little ammonium phosphate is present various kinds of lice (plant-lice, root-lice, phylloxera &c) are produced.

If air is absent we get the marine phosphorescent sea-nettles.

If common salt and limestone attach themselves to the nerve-albumen, there originate, if air is lacking, the clumsy crustaceans and the oysters with weak senses.

Even among the mammalia similar phenomena are observable. The massive bony ox while being harnessed to the plough, receives many a blow from the ploughman ere he learns his proper place. How different the intelligent dog with his comparatively light-built bony structure!

The noble steed too—the heavier his bones the less mettle does he show. What a difference between the dray-horse and the percheron!

No otherwise is it with man. It is hardly necessary in this connection to contrast the spindle-legged delicate figure of Voltaire with that of a coarse-boned quarry-man.

It may in general be said that intellectual life appears the more prominent—the finer the nutriment which is absorbed. Our fly-eating singing birds, our bug-eating nightingales which as long as the sun shines seem mere embodied song, and while they live and breathe are nought but movement, joy and happiness—they show us what kind of material lecitin is. And the sugar-produced and flower-sugar eating bees and butterflies, do they not show that life, movement and happiness are dependent on the oxidation of saccharine matter in the rays of the sun!

The nerve-stearin-fat (lecitin) in combination with saccharine matter—"protagon"—is the basis of innumerable different kinds of albumen, in so far as sugar is capable of combining in various modifications with earths and salts. For example lime can both alone and in combination with phosphoric acid, sulphuric acid, the halogen acids, and hydrochloric acid enter into a chemical condensation with sugar. Similarly potash, soda, magnesia, manganese and iron. This condition of things makes it clear that the proportions in which the various chemical constituents of the different albumens are mixed together form the basis for the accumulation or aggregation of similar material so as to give rise to a variety of vegetable and animal forms, as I have already pointed out in detail in my book "Das Leben".

A fixed relation between the chemical constituents must necessarily exercise an influence on the further development, and the absence of a particular substance must modify the character of the resulting animal or vegetable organism. I might give as examples the vine and the currant bush, which both in leaves, fruit and the ashes of the seeds, possess such a great degree of similarity. In the currants however chlorine is wanting, while it is present in grapes, and this circumstance gives rise to a difference which must necessarily find expression in the form of the plant.

As a logical consequence of this we observe, that the shapes of both plants and animals alter when they begin to be deprived of the substances which their peculiarities render necessary to them. To what an extent does not the form of a consumptive change!

What fungoid growths do not leaves and twigs degenerate into?

The occurrences of such degenerations lead us to the conclusion that every particular shape requires an adequate supply of those substances which determine its form.

As far as the mammalia are concerned they are, so to speak, double creatures which owe their origin to different though closely related kinds of albumen, viz the nerve-albumen (lecitin) and the blood-albumen with a varying hemoglobin basis. The horse, the sheep, the ox, the dog, the cat, the pig, the hedgehog, yield a hemoglobin crystalizing in a special form; sometimes rhombic, then again the rhombus with truncated corners as hexagons, and these hexagons, in variously elongated forms.

Human hemoglobin crystallizes in regular equilateral hexagons.

Nerve-albumen, and blood-albumen accompany each other in such a way that the tubes in which each is inclosed run parallel to one another to the apices of both; and it falls to the lot of the arteries to carry the vivifying oxygen to the oil-saturated nerve-fibrils, while the venous system in a parallel course carries away the products of the oxidation of the nerve fat.

In as far as our bodily substance depends upon two different though allied kinds of albumen which belong together, it is plain that a disturbance can take place in the vital processes owing to a derangement of either of these systems, but it is also clear that either of these systems may be acted on through the other i. e. the nervous system through the blood or the blood through the nervous system. I only make this remark at this point "*en passant.*" Unless we are informed as to the materials from which our body is built up, we are not in a position to study its organic structure, as we must continually take our stand on the chemical basis.

THE BODILY STRUCTURE.

In the egg, as soon as an electric impulse has been produced by the coming together of nerve-albumen and blood-albumen, the originally homogeneous substance rearranges itself into groups of opposite constitution. On the one side a chemical separation of water takes place accompanied by a corresponding condensation or solidification, while the water which has been set free liquifies an other portion. In this manner originate numerous ray-like strings, filaments or fibrous matter, and on the other side a fluid lymph which is enriched with the salts thrown off by the fibrous matter on its separation. These salts give to the lymph a higher electrical tension, for all salts excite electricity and the more energetically the greater their degree of concentration. But in addition to the lymph, and the fibrin which has been formed, there is also a certain amount of unaltered albumen left, which can supply the necessary material for further development in both directions.

Fibrin, albumen, and lymph, these three substances originating from the albuminous matter, supply us with a key to the structural organization of our body, including the bones, for it must be remembered that bones are nothing but cartilaginous substance within the tissues of which phosphate of lime has been imbedded. On dissolving out the lime with diluted hydrochloric acid, we obtain an elastic mass of cartilage preserving the original shape of the bone. It must therefore be understood that wherever the expression "bone" occurs we have before us "cartilaginous substance." I will now give examples to show that this three-fold division into nerve-substance, fluid lymph, and hard tissues (bone, cartilage and fibrin) is of universal application to our whole bodily organism.

In the first place our specific nerve-centre the brain-albumen may be divided into the filamentous dura-mater, and into the arachnoid membrane between which is enclosed the serous cerebro-spinal fluid, the whole being surrounded by the protective bony cranium. In the interior we have the albuminous cerebral matter as a productive base.

We must regard the cortical stratum of the brain-albumen, the grey matter, as radiating electricity like a conductor emitting sparks. It is continually oxidizing itself by the oxygen of the blood to form carbonic acid, water, nitrogen and phosphate of ammonia, and is constantly renewed as it is used up like the mucous membrane of the stomach which is dissolved into pepsin &c. No organic substance can produce warmth or motion or do any other work without at the same time being chemically dissolved. In fact this is demanded by the law of equivalence of force. The fat of the sweat-glands and of the eyes, as also the wax of the ears are to a certain extent the equivalents of the activity of our senses.

To the same extent as the grey brain-material is used up, it is redintegrated from the subjacent albuminous brain substance, which is in turn renewed from the albumen of the blood supplied unceasingly by the basilar artery.

The general relationship of the brain albumen within the cranium is preserved outside in the various branches or processes known as the cerebral and spinal nerves.

The first of these branches which issues from the brain is the *olfactory nerve* which divides its material into the Pituitary-membrane and into. fibrous *connective tissue* and *cartilage.*

The *optic nerve* divides into the *horny* cornea, the *fibrous* sclerotic, the retina and the *lymphatic* vitreous humour on the one hand, and into the lacrimal *glands* formed of mucous tissue, into lymph-like *tears* and the *eyelids* of connective tissue on the other. (When we weep, we weep away a portion of our optic nerve as an equivalent for our mental pain). In the connective tissue of the eye-lids we have the same division: the tarsal cartilages, the ciliary muscles, and the Meibomian glands with their fatty secretion bear witness to this standing rule.

The case is the same with the *auditory nerve.* It is uncertain whether the auditory or the optic nerve deserves the palm. I will merely draw attention to the *cartilaginous pinna* or auricle, the passage lined with *epithelium* and the fatty *lymph* known as the wax, and within, the cavity of the tympanum lined with *mucous membrane* and the *fibrous* tympanum itself. The tympanic membrane shows connective tissue, mucous membrane, and lymph, also the Eustachian tubes have bones, cartilage epithelium and mucous glands. The walls of the semicircular canals too are provided with bones and cartilage, and their enlargements have connective tissue and a transparent serous membrane in the cavities of which is the so-called endo-lymph, and in the pinna or auricle serous periost and perilymph. Finally the cochlea shows the same subdivision—the *fibrous* organ of Corti the accompanying *lymphatics,* and the *osseous* laminae are all of them connected parts of that portion of the brain marrow which has been transformed into the auditory nerve.

I will only point out in few words the most important portion of the brain—the tenth pair of nerves, the conversion of which into cartilage, connective tissue and epithelium bears strong testimony. We have the wonderful arrangement of muscle and cartilage known as the larynx, the partly cartilaginous—partly membranous bronchi, and the thymus-gland conveying lymph, and which in the adult is absorbed by the pleura, and represents the third subdivision with the cartilage and membrane. It is suspended between the trachea and the sternum like the lacrymal glands between the eye and the eye-lid. As pendant to the larynx, the trachea and the thymus gland there again occurs in the lungs and stomach an elastic expansive fibrous tissue together with a glandulous mucous membrane and lymph-containing membranes (Pleura and Peritoneum).

Similarly to these processes of the brain which are converted into face, lungs, oesophagus, and stomach, is it with the stem and branches, with which we are acquainted under the name of nerves of the spinal medulla and the medulla oblongata.

It is sufficient to draw attention to the spinal column with its tendons and lymph canal, to the cartilages of the ribs with their tendinous connections, the lumbar nerves and the sacral plexuses and ligaments, the sciatic nerve with the bones and ligaments of, the leg, and on the other hand to the lymphatic region of the inguinal glands, the peculiar testicles, and the extensive system of the lymph arteries of the arms and legs.

Everywhere we have tendinous fibre yielding glue, the mucous membrane of the connective tissue, and lymph.

This arrangement is most apparent in the *eye* where the optic nerve abundantly supplied with blood rich in albumen spreads out spherically to form the sclerotic coat and the retina, enclosing at the same time the so-called vitreous humour, consisting almost entirely of lymph, while the horny and hard crystalline lens containing globuline-albumen, and the gelatinous cornea complete the whole. The lymph of the vitreous humour contains such a very small amount of gelatine, that the delicate membrane formed therefrom which constitutes a sort of series of scanty partitions for the lymph, even if very slightly injured, permits it to escape. As the retina upon which the rays of light fall after passing through the crystalline lens, is woven of nerve substance which is liable to be consumed, the lymph of the vitreous humour seems to be the substance from which it is recruited. The anatomical aspect of the case likewise renders it probable that the lymph has the mission of nourishing the nerves; for the lymphatic system sends out innumerable branches to the nerves, and there can be little doubt that the fatty contents of the lymph penetrating through the gelatinous membranes of the nerve trunks serves to nourish and renew the nerve-substance. This operation is rendered easier by the very slight amount of gelatine in the lymph, in which respect it presents a resemblance to the nerve substance itself, which contains only one proportion of "gelatine-sugar" with two of carbon, as against a mass of fat generated from saccharine matter with 40 parts of carbon. The delicate nature of the Pia Mater and the internal cerebral membranes or septa is in accordance with this small content of gelatine. The chemical relationship of the nerve substance to the oleaginous lymph is patent. In addition however we have the fact that only a small number of blood vessels, but a large proportion of lymphatics are connected with the nerve trunks. There is consequently a sufficient ground for the conclusion that not only the nerve substance is recruited from the lymph as it is consumed, but also that considering the contrast between the lymph and the muscular tissue, that it is from the lymph that the soft oleaginous nerve substance was originally formed.

In fact the nerve-nourishing mobile lymph is to be regarded as a refined product in comparison with the firm substance of the tendinous tissue with its relative chemical unalterability and great capacity for resisting decay and corruption. While the tendinous tissue is relatively speaking passive, the nerve material sprung from the lymph, owing to its oxidizability is really to be regarded as the dynamic source from which all the various impulses and motions originate. Its main constituent as already explained is ordinary tallow-fat. This fat experiences a greater degree of chemical combustion during the day than during the night, when its substance is more especially undergoing replenishment from the lymphatics. During the day, therefore, it not only produces light and heat internally, but is of such importance as the driving power for all the various apparently mechanical processes of the body, that we can easily under hostandy it is, that it depends upon the abundant supply of nerve substance, whether we set the body in motion and continue in motion, or whether our exhausted frame shall fall to the earth in accordance with the law of gravitation.

It is the nerve force which sustains the body erect.

Albumen, Fibrin, and Lymph! The occurrence of fibrin and lymph which originate from albumen, is the result of a chemical movement of the atoms which was started by the meeting together of two kinds of albumen of similar but not altogether identical chemical composition. The movement thus initiated continues, new material beginning to take a more and more pronounced part in the various phenomena of motion and of life.

From the albumen supplied in the case of the mammals by the blood of the mother, a continuous chemical separation takes place of solid and liquid constituents giving rise to muscular tissue and lymph.

A local condensation and a corresponding rarefaction accompanies each one of the undulations of heat attending each wave of blood, and each pulsation of blood which the heart of the mother propels through the foetus. This fact is best illustrated by a mechanical experiment If 8 to 10 small marble balls be placed like pearls on a string, in the depression along the rim of a plate, and a blow be delivered with a knife to the last one, two or three balls will leave the other end of the row, more or fewer according to the force employed. In a similar manner under the influence of the maternal life there is detached from the albumen of the blood flowing into the new organism a certain amount of carbonic acid and water, with an accompanying conversion of the grape-sugar of the albumen into olein, and of the gelatine sugar to muscular tissue. This process occurs in a similar manner to the way in which it takes place in plants as an effect of the electric rays of the sunlight and of the heat of the sun.

As far as the fibrin is concerned, which is produced from the blood-albumen, it may be shown that nothing further is needed for its formation than a blow or impact together with admission of oxygen. If air be forced into blood which has been let from a vein, by beating it with a stick, the fibrin of the blood adheres to the stick in the form of satin-like threads. A similar occurrence takes place inside the organism; wherever it encounters anything already solid, there it remains and solidifies. Thus it attaches itself to the little crystals of phosphate of lime, which are carried along by the all-dissolving blood, rich in salts and crystallize from solution in it, producing on the one hand cartilage and bone and on the other the fibrous tendinous tissue which penetrates the substance of the bone to a certain depth. It thus becomes either a firm connecting bridge between two so-called "bones", or it presents the arm of a lever to which the muscular bundles which arise from the albumen of the blood are attached. Thus bone and tendon, and tendon and muscle pass into one another by insensible degrees.

The bands or fasciae so hard to rend, which when wet shine like satin are in reality "satin" or silk but satin in its strongest most compact and solid condition. From this satin-stuff are formed the extraordinarily powerful "synovial sheaths" which seem as if forged on to the bones of the wrist but which are in reality woven into the bony substance. It is underneath the arch formed by them that the tendons pass from the arms to the fingers.

The gelatinous silk of our tendons is formed by oxidation from the gelatine-albumen of the blood in quite a similar way to that in which ordinary silk is formed from the albumen of the silk-worm when the albumen passes out from the spinning glands, into the oxidizing air water and carbonic acid being separated at the same time.

The fact that the gelatine of the cartilages on boiling in hydrochloric acid takes up water and yields grape-sugar, renders it certain that it must have been formed from dehydrated saccharine matter and ammonia; and this fact also explains the formation of sugar in the liver from the plasma of the blood during diabetes when mental irritation occasions a chemical splitting up of the blood albumen; lymph and fibrin being formed at the same time.

Further the true nature of albumen as depending on grape-sugar and gelatine-sugar, and of fibrin as mainly containing gelatine-sugar is best studied in the silk-fibre.

Raw silk contains besides two thirds of fibrous substance (Fibroin) one third of gelatine (Sericin), and these two substances manifest themselves as decomposition products of the same original material.

Sericin C_{15} H_{25} N_5 $O_8=5$ of gelatine-sugar plus 1 of grape-sugar plus 2 of oxygen minus 6 of water and 1 of carbonic acid. The 2 of oxygen are withdrawn from 2 parts of Fibroin for

Fibroin C_{15} H_{23} N_5 $O_6=5$ of gelatine-sugar plus 1 of grape-sugar minus 7 of water 1 of carbonic acid and 1 of oxygen.

Observations of this kind made upon the chemical decompositions of silk-fibre spread a flood of light upon the origin of animal tissue and its alterations during disease. The silk of the silk-worm is, however, by no means free from ash constituents, on the contrary it always contains at least $^6/_{10}\,^0/_0$. This content of earthy substances is conditioned by the chemical combining power of the gelatine-sugar which forms the basis of the albumen. The gelatine-sugar COO, CHH, NHHH, owing to its content of glycollic acid (COO, CHH) and of ammonia (NHHH) is enabled to combine with both acids and bases. This fact completely upsets the view that inorganic substances (e. g. salt, glaubers salt, sulphate of lime &c) are incapable of being assimilated by our organism. which in addition is controverted by the salutary effect of the saline mineral waters. Further such mineral substances as are themselves insoluble in water (e. g. phosphate of lime) become soluble by combining with the gelatine-sugar of the blood, or the albumen of the lymph. This is attributable to the hydrocarbons (CHH), contained in the gelatine-sugar. What its effect is we can understand from the solubility of the methylic sulphate of baryta as compared with the insolubility of sulphate of baryta. From this we can understand very well the way in which the lymph containing sugar and gelatine-sugar can transport in solution earthy substances like phosphate of lime to the parts where they are required, and how this phosphate may be deposited when either acids (e. g. lactic or oleic acids) combine with the gelatine-sugar, or when the hydrocarbon CHH which serves to keep it in solution is oxidized by the oxygen of the blood to carbonic acid COO and water (HHO), which is carried off in the venous blood.

We consequently know at any rate this much concerning the nature of fibrin, namely that it is characterized by its containing a greater amount of "gelatine-sugar"* and by the chemical detachment from it of water, and that its origin from albumen is due to internal chemical motion, and it is plain that the dehydrated condensed, and non-sensitive tendinous fibre which resists putrefaction, draws further material out of the albumen according to the extent to which it is subjected to tension. Thus we observe a notable development of fibrous tissue in the case of hardworking men and women, and in draft animals, and this takes

---

* Glycocoll.

place at the expense of the muscular tissue by the albumen splitting up into its fluid and solid constituents. In this regular process we find also the explanation of the natural hardening of the tissues in old age. This hardening may reach the condition of ossification, and by the exhaustion of the movable materials finally causes death from old age.

The regularity of the process by which a pull or blow, a push or pressure causes a chemical condensation, which requires as equivalent on the other hand a separation, the result of which appears in the form of chemically detached water and carbonic acid, is seen not only in the palm of the hand, but also and, indeed, in striking beauty in the ground of silk, shining like satin (the fascia), inside of the connective tissue of the sole of the foot; with these again the strong tendons connected with the muscles of the thighs are in close relation. This silky tissue of the fasciae of the soles is about the most beautiful thing that the dissecting scalpel reveals to our admiring eye; and we understand from examining it how the baby who is allowed to stretch his feet against the crib or wave his little arms, puts tension on his fascial tissue, developes it and so as it were spins silk; an advantage which is denied to the child bound up in swaddling clothes. The peculiar twisting of the feet which characterizes the little-ones, is a case of spinning silk for the development of muscular strength; and yawning too is the same thing as regards the ligaments and muscles of the jaw. Similarly freedom of movement of the spinal column in children conduces to the equal development of the ligaments which connect one vertebra with another, a fact which may be observed while carving the back of a hare or deer. In opposition to this we find the ever more frequently occurring mal-formations of the spine to be due to the want of freedom to which some children are condemned with the good intention of giving them an erect carriage, but which in reality causes impediments in the circulation, exuberances in the cartilages, and inflammations of the lymphatic glands.

Having now made some acquaintance with the nature of fibrin, we will proceed to the consideration of the lymph.

Hitherto the contents of the lymphatics have been regarded as to a large extent consisting of waste or excretory material. Among others Jos. Hyrtl says in his "Lehrbuch der Anatomie des Menschen" (Vienna 1878): "The physiological function of the lymphatics is directed to the absorption of the liquid constituents of the blood after they have passed out of the capillaries and have done their work of building up the tissues, and to their restoration to the circulation."—This, however, is the function of that portion of the capillaries with which the veins commence. The removal of the waste or decayed products is provided for in the system by the highway of the extensive intestines and capacious bladder. Finally the principal lymphatic which runs parallel to the spinal column—the thoracic duct—enters the circulation quite close to the right auricle of the heart where the internal jugular vein and the subclavian vein unite. Now as it is plain that it is the function of the heart to circulate through the system the nobler kinds of blood, waste and decayed products have no business to enter there.

In direct opposition to the view that the lymph consists of waste products, we must on the contrary regard it as representing the most noble material with which the organism is supplied.

By the decomposition of the albumen, which originally contained grape-sugar and "gelatine-sugar" in equal proportions, into fibrous tissue and lymph, a large quantity of "gelatine-sugar" combines with a small proportion of grape-sugar, giving rise to a separation of water and carbonic acid as we saw in the case of silk. The lymph presents a contrast to the muscular tissue in containing but little "gelatine-sugar", that might curdle into fibrin, but it contains in addition to the albumen, the sugar-material and the earths and salts that were not made use of for the production of fibrous tissue. It is its richness in sugar that enables the lymph which flows in its own particular channels out of reach of the oxygen, to supply the materials required for the formation of fat, and this takes place in such a way that carbonic acid is separated from the sugar. In all probability the carbonic acid thus liberated assists the forward movement of the lymph, which takes place in fine channels with a construction similar to that of the locks on a canal; and since it has to supply the whole organism, its upward movement is facilitated by the presence of a number of "relay-stations"—the lymphatic glands—from which it is sent on further. Through its contents in ammoniacal albumen, phosphatic earths and salts, saccharine matter and fats, the lymph possesses all the nutritrive elements both for the blood and the nerves, both of which have to obtain therefrom the materials required to supply their waste.

If then the lymphatics with their "relay-stations" the lymphatic glands, have this important mission to fulfil of supplying fresh material to the blood and nervous system, it is to be hoped that those gentlemen who have hitherto been so ready to remove swollen lymphatic glands with the knife, because it was customary to say that no one really knew what was their use, will in future be a little more backward in making operations.

The lymphatic-system is like a network of railways with junctions, having roads entering these junctions, or if we choose, issuing therefrom like rays; nobody can say, where it begins or where it stops. The commencement is really every where, as it should be with an intermediate factor between blood-material and nerve-substance.

Returning to the point of view from which we set out, that it is the nervous system which affords the best clue to the comprehension of the anatomy of the whole body, we find that the brain-substance which is protected from immediate contact from without by membranes of different thicknesses and by a bony-shell of gelatine and phosphate of lime (the cranium), consists essentially of a chemical compound of saccharine matter with fat and a phosphate (ammonium phosphate). It appears to be divided by a number of delicate membranes into compartments each of which is capable of independent development like the shoots which sprout from the "eyes" of a potato. To them correspond the various nerves which protected from immediate contact by membranes and in places by osseous walls, issue symmetrically from the two halves of the brain, in the one case creating in front the face (facial nerve), in the other branching off behind to help to form the lungs, stomach (vagus), and this in such a manner, that the branches of the nerve trunks from which the lungs and the stomach arise, continue on so as to reach the liver, spleen, pancreas, and kidneys.

To afford these organs support, the globular brain substance is prolonged into the spinal-cord protected by the vertebrae.

The spinal nerve trunk too gives off various branches provided with a gelatinous sheath (perineurium), among which we may enumerate the cervical, brachial, intercostal, lumbar, and sciatic nerves.

Thus we may recognize the following pairs of processes from the brain:

1. Olfactory.	7. Facial.
2. Optic.	8. Auditory.
3. Motores oculi.	9. Glosso-pharyngeal.
4. Pathetic.	10. Pneumo-gastric.
5. Trifacial.	11. Hypo-glossal.
6. Abducens.	12. Spinal accessory.

The three latter grouped in each of their halves closely together, surround to a certain extent the stem-like prolongation of the brain known as the spinal cord, which in turn is surrounded and protected by the hard vertebrae from which issue 32 pairs of nerves — 8 for the cervical plexus, 12 for the thoracic and 12 for the abdomen and legs.

These various prolongations of the nerve substance of the brain form together the really movable portion of our bodies. Their movements would be able to take place in all directions, like those of rainworms, were they not confined, like the mediaeval knights were by their armour, so they by the surrounding bony formations, which prevent motion in certain directions.

As far as these bony parts are concerned they may be regarded chemically as a conversion-product of the brain-substance. The bone contains fat and ammoniacal gelatine, just as the nerve substance does; the only difference is that the nerve substance contains phosphate of ammonia and the bone, phosphate of lime as the chemically connective material.

Indeed when the lime is exchanged for ammonia in the bone gelatine, we find the bone converted into living nerve material which commences to move, as is shown by the maggots which originate in ham bones. Thus the bones are in reality nothing but masked nerve-substance.

Such being the condition of affairs it is but little to be wondered at, that we find a second nervous system arising from the bony spinal column. For there is supended, attached by different fibres at various places—the so called sympathetic system—the branches of which in conjunction with those from the cranial and spinal nerves assist in forming the intestines. This intestinal nervous system is a net-work of plexuses unlike the brain processes which run in comparatively straight lines. In consequence of its passage through the petrified form of nerve-substance (bone) it is so to speak "polarized" i. e. has lost the power of independent sensation. At the same time, however, its fine terminations regain the power of sensation through being blended with the last most delicate fibres from the cranial nerves to form plexuses. The cementing material between the two kinds of nerve fibres is on the one hand muscular tissue and on the other, connective tissue. In this way the electric current dwelling in the nerve fibres is compelled to form a closed circuit, in consequence whereof under ordinary circumstances it is not sensible to the subject. It only makes itself perceptible, when an interruption takes place in the regular current of the contents of the veins or arteries running parallel to the nerve fibres. We always find that the blood-material and the nerves form complements one to the other; and the fact that the blood vessels are penetrated by spirally arranged fibres of the sympathetic nerve, permits us to regard them as counterparts to the cerebro-spinal branches.

In so far as every nerve fibre is accompanied by blood vessels which are frequently enclosed in a common sheath, the unity of the structural plan of the human body becomes apparent.

Consider the case of a stearine-candle! When it has been lighted we can by its flame melt tin in a spoon, or boil water. We have then to acknowledge in the molten condition of the metal or the boiling state of the water the result of the expenditure of energy on the part of the candle.

If instead of burning the stearine of the candle at a single point we could conduct it in its melted state through a system of branching tubes and wicks, and light it at once in a number of places, we should be able to obtain a multiple effect at a variety of points. And if care should be taken by the insertion of a proper system of pipes in the heating apparatus to constantly supply fresh fluid stearine, such an apparatus might be kept working for a hundred years, or more yet without interruption, if no injury were done to the mechanical portions by unequal heating or by external force.

Such a machine is Man!

Our vital force depends on an uninterrupted burning of nerve-stearine, which owing to an electric process has been brought to a state of tension, after which it has attracted further nerve-stearine to itself, the first oxidation product of which in the form of gelatine supplied membranes and protecting tubes.

As common oil burns in the wick of a lamp, so does our nerve-oil burn in the fine ramifications of the wick-like nerve-fibres, by means of the oxygen which the arteries supply which arise from the finest capillaries of the lungs.

Owing to the fact that the nerve-oil forms a chemical combination with oxygen throwing off minute portions of water and carbonic acid as products of combustion, the process of combustion in accordance with the law of the conservation of energy, is equivalent to the operation of a vast number of small portions of energy, which in their perpetual succession, make up the total or "capital" of our vital force. Our vital force is thus essentially dependent upon the continuity of the process of combustion. If the combustion be interrupted, as for instance by suffocation or drowning, the matter comes to an end. The secret of the vital phenomena lies in their momentaneity. The vital force is born afresh every instant in virtue of the act of respiration. When the respiration stops the vital force disappears. That is the point of which we must keep a firm hold! It is the central point from which everything may be reached and understood like the various threads of a spider's web from its central point.

Respiration takes place in the lungs by means of the hair-like blood vessels with which it is supplied, which present an anatomically inextricable "ant-hill" the functions of which, however, take place with just as regular a plan as in the case of the ant-hill.

The lung-capillaries unite together to form larger vessels which at length form a large receptacle—the heart.

The function of the heart, which is set in motion by the electric force of its nerves, is similar to that of the water-chamber of a steam fire-engine the contents of which are forced forward through tubes. In this way the lung- or arterial blood reaches the remotest parts of the organism and there ignites the melting stearine-fat of the nerve-points by means of the oxygen carried from the lungs. This it is which forms the chief source of power in air-breathing animals, whereas in aquatic animals the separation of nerve-substance partly in the form of water, partly in that of carbonic acid represents the principal form which the development of force assumes, and is analogous to the violent convulsions of epileptics and of those who are hung, which are likewise due to insufficiency of oxygen and are based on the separation of nerve substance, just as the liberation of mucus is followed by convulsive conghing as an abnormal motion.

In air-breathing man the regular manner in which force is developed, depends on the constant combustion of nerve-oil by means of oxygen—the result being that every portion of the nervous system to which no oxygen is supplied becomes debilitated and liable to chemical decomposition or to decay.

Since however our vital force can only become active through the combustion of nerve-oil, it is plain that fresh nerve-oil must be periodically procured in order that the lamp of life may continue to burn.

Thus a complete correspondence prevails between the consumption of oil in an ordinary lamp-wick and the burning of nerve-oil in the wick-like ramifications of the nerve fibres, the only difference being that the combustion of nerve-oil does not occur in a large free space but on the contrary in a confined narrow space which gives rise to a certain amount of difficulty in the access of the air through the pores of the capillaries to the nerve terminations. We have, however, the advantage that the vital flame being spread equally over millions of different points, no general conflagration can occur, but the combustion has the effect of producing a mild warmth of which (as not much is lost by radiation) the body receives nearly the full benefit.

The difficulty of conveying the oxygen to the nerve endings is increased by the circumstance that the channels by which the products of combustion are carried away from the terminations of the nerves, are just as narrow as those through which the oxygen is introduced, and instead of running in straight lines to the surface, they get there through many convolutions. It is however this very point which enables complete and full use to be made of the warmth resulting from the heated products of combustion—water and carbonic acid—which pass by a circuitous route to the lungs where they are set free—the "ant hill" of lung-capillaries having meanwhile its hands full of work.

Since however the products of combustion of the nerve-stearine carry with them the products of combustion of the partitions, their ash-forming constituents also, as giving rise to salts must be considered. This also in fact takes place, thanks to the lung-capillaries, with such order that the solid salt-particles are carried along tubes with strongly resisting walls to still more resisting glands—the kidneys—by means of branches from the abdominal aorta where they, dissolved in the serum, are excreted in the form of urine.

This means of distributing the products of combustion—the gaseous ones going upwards—the solid ones downwards—assures the continuous and abundant combustion of the nerve-oil in so far as an accumulation of the products of combustion in the lung capillaries is not occasioned by ignorant omission of the act of expiration, which would mechanically obstruct the entry of fresh air and so cause a diminution of vital force at those points, giving rise to a disease of the organs and a disorderly disintegration of nerve substance.

When we consider our whole anatomy from the above points of view, we see that the following vessels or systems of tubes must be present:

3*

1) A compartment for the nerve-stearine. This is the cranium and the canal of the spinal column with their bony and membranous prolongations.

2) A compartment for the blood carrying oxygen. This is the left auricle and ventricle of the heart with the aorta and the arterial system.

3) A compartment for the watery blood containing carbonic acid (venous blood). This is the right auricle and ventricle of the heart, the *vena cava,* and the venous system generally.

4) A system of tubes for supplying fresh nerve-oil. This is the lymphatic system which enlarges itself at last into intestines and stomach, into which latter fresh material is introduced through the œsophagus and the mouth-funnel.

5) A system of canals for alternately introducing fresh air for combustion and carrying off the burnt gases. These are the lungs with the bronchial tubes or in other words the respiratory system.

6) A system through which the heavy ash-forming matters and used up salts are got rid of. This is the large intestine on the one hand, and on the other the kidneys, ureters, bladder, and urethra (excretory system).

The walls of the different systems of canals running at times parallel to one another, at times interwoven in their ramifications, make up a harmonious whole in the form of a pliable mechanism. Since further the different systems run one into the other, any derangement of one must necessarily make itself felt by the whole. Hence it is meaningless to say that a man suffers from the liver, for it is certain that in such a case a number of other points must also be involved, as we have before us an interdependent piece of clock-work.

Considering everything together, we come to the conclusion that the mechanism of our organism is essentially composed of combustible nerve-oil and is sustained by the combustion of this nerve-oil. The principal thing, therefore, is to prevent interruptions in this process, or to remove them when they occur, so that the vital flame may continue to burn. We will consider the conditions under which this is attainable, in the next section.

II.

PHYSIOLOGICAL PART

OR

THE ACTION OF THE BODY IN HEALTH.

"Out of the tools and machinery which he has
created let man rise up before himself as a "Deus
ex machina".
Dr. E. KAPP, Philosophie der Technik.

All the various functions of the body without exception have for their end to render possible our mental activity. The circulation of the blood, respiration, digestion, excretion—all tend to enable the nervous system to exercise and continue in the exercise of its functions.

Mental activity in reality, however, merely consists in assembling together bodily material. By means of thought and meditation we impart motion to the materials around us, combine them in groups and shapes and derive pleasure from the manifold forms resulting therefrom. It is in this that our divine inheritance consists; "know ye not what manner of spirit ye are of?"

"Consider it well. The inference lies so close. You yourselves are God become Man. Recognize him in yourselves and he is there." (Sallet).

To build, to create, to gather together, to rule the things around us that is the purpose for which we exist. How could it be otherwise? Everything that exists is from God, and through God, is God himself, who permeates all matter. As a consequence we are ourselves a portion of God, and cannot otherwise be happy in our existence except by striving to resemble God in creative activity. There is in this matter a natural compelling impulse. As God without ceasing new-creates His creation by transforming it, as He directs and rules it, so consciously or unconsciously we do the same. How extraordinarily far from re- cognizing this fact are the hitherto existing works on physiology which supply a profusion of separate facts without showing the principle which unites them into a whole. The consumption of tissue, the circulation of the blood, pulsation, respiration, muscular movements and sensations are presented to us in rigid dissection.

> Philosophers and sages, men of might
> They call the searchers, who to learn truth's way,
> Break down the stately organ fair and bright,
> While Music pines and dies away.
>
> If they but know enough to call by name
> Each nerve and artery and every tissue
> They think they've merited earth's highest fame,
> Forgetting that 'tis Life, which is the issue.

(Sallet.)

It is with a shyness almost amounting to fear that existing Physiology has stopped before the very kernel of our being *i. e.* the action of the soul, and has only busied itself with the case and has been highly delighted whenever it could succeed in pointing out the portion of the brain which corresponded with any particular exercise of mental activity. Matters of quite subordinate importance and completely outside the sphere of mental action e. g. the order of movement of a horse's feet have brought the title of "Master of Physiology" to whoever occupied himself therewith, and the imitation of the human voice by means of a piano-like mechanism even procured a pension from king Louis of Bavaria.

It is upon a constant re-forming and carrying away of the former material that the whole of the functions of our body turns and hinges, while at the same time the introduction of fresh homogeneous substance of more powerful and resisting structure as far as possible, takes place for the purpose of enabling mental activity to continue and this action consists properly of motion and change. We observe the childish views and affections alter in proportion as the child's body and the materials composing it give place to a more mature form and to other and firmer substance. What is novel and hitherto unknown coming in contact with the mind brings it fresh nourishment which is assimilated causing increase and enlargement of the brain-substance, and it is only when it is no longer affected by what is new, that what we in contrast with "Life" term "Decay" commences. Nevertheless this apparent death is in reality nothing but a radical rejuvenescence, and reformation.

> "Every death is but a resurrection up above
> The eternal escape of the spirit to God."

Sallet.

By far the greater portion of our lives is taken up by the activity of thought, whereas e. g. eating and drinking only demand a relatively short time. The mind is in truth far more desirous of food than the body, so that for want of better things a great number of men find enjoyment in the most worthless chatter, while with others the impulse to mental exertion when it is not satisfied from without by the changing impressions of things, rests on itself, i. e. turns inwards upon itself and becomes inventively productive. Minds of this nature gifted with the divine power of imagination or fancy do not as a rule stand high in the estimation of the more dense witted portions of mankind. While still dwelling in the realm of the dull-witted they generally go by the contemptuous name of "dreamers or phantasts", and it is only when they have shuffled off this mortal coil that men would strive to retain them and imitating their form to preserve it for posterity in monuments of stone and bronze.

But nevertheless, men gifted with imagination influence a portion of their contemporaries like fermenting leaven. When they speak, others cease to discuss the trivial concerns of daily life. That this is so, is due to the presence of a divine spark which slumbers in every human breast, and which only requires a breath to blow it into flame. Well is it for us when we come within the magic circle of spirits like these —we awaken then from weariness and slumber.

"How steel strikes upon flint! How sparks are scattered! Thoughts bud forth from thoughts full of productivity and set all eyes aglow. Here must (the spirit knows no limit) Man complete himself to divinity. Make the attempt and you will be internally illuminated by God, and exult as the lark in the breath of morning. The gray twilight-world must then be bathed in the rosy glow of thought. What you have sought so far away, it is so near! Thousands of years, and millions of miles disappear into nothingness, for God is there. He dwells in the spirit upon an invisible throne. God was far from you: you beheld in the universe nothing but machine-like movement of dead matter. God lives in you; in movement, in brightness and in sound, you may behold the eternal forward motion of the spirit. But even if you lay in the darkest dungeon's night, even in the dungeon you might build up the kingdom of God. Call him and he will come into the depths of your heart, and make divine both will and thought and sight. Why do you point gloomily beyond the grave? Not complaint, but sturdy labor should be your hope. Thou art in God, Thou wast in him from the beginning, and here as there heaven stands open to thee." (Sallet).

THE BODILY ACTIVITY DURING WAKEFULNESS.

The mystery of sensation consists in the fact, that a particular excitation is transferred to us and causes the mind to move in a corresponding manner. The sensation of sympathy or of pity is in this respect the mother of a number of varieties of emotions, and sympathy proves kinship of mind. I consequently quite find it impossible to regard individuals destitute of sympathy or pity as equals or as brothers, but withdraw from them as much as possible, and treat them as they deserve, that is to say as strangers, with distant and chilling demeanour. It is in this respect plain that the chemical composition of the material which became active in the formation of the original germ is of the greatest importance: the shark cannot be blamed for his unenviable peculiarities nor the philosopher praised for what he produces; only no one should demand that a philosopher should walk about amicably arm in arm either with a shark or a grave-robbing hyaena.

As it cannot be denied that almost all classes of animal characters are represented among mankind from the ape to the worm, and since sympathy refuses to be moved except by what is like or homogeneous, it follows that we are justified in averting ourselves from what is contradictory and opposed to our nature as much as possible, and even to subdue, and to tame and rule it.

To attain to this power of command the principal requirement is mental superiority—mental force. What gives rise to this mental energy?

Since the far reaching discovery of Dr. Robert Mayer, the practical physician of Heilbronn, who was the first to express the equivalent of force numerically, we are aware that there is in reality but one fundamental force in nature, and it is unimportant what name we give to it. For my part I am disposed to term it "contrariety" or "polarity" since it is an opposing or dissimilar nature which successively causes attraction, motion, impact, heat, electricity, magnetism followed again in its cycle by attraction. Since, however, this expression is not as yet generally current, it is advisable in order to be generally intelligible to make use of a well known word. We will therefore say that the original force is electricity, even though we can merely regard this as a stop gap, for it is quite certain that electricity is only a partial force and is in no way the whole force. Anyways it is universally known, that electricity when retained in bodies by suitable arrangement is capable of producing light and heat, movement and sound, as also magnetism and chemical combinations, while by dividing up electricity into negative and positive halves we are enabled to separate chemical compounds from one another.

There is no doubt that every species of movement with which we are acquainted as occuring in the world can be traced back to electricity, and even our vital movements draw upon this common source. A few examples may give the required degree of clearness in this matter.

A brook flows down the valley and turns a mill. That is electricity! — How so?

The connection is certainly not to be discovered by the microscope. On the contrary instead of shutting only one eye as when using the microscope, we must rather shut both, and turn our mental vision toward the sun. There we find the chemical elements in gaseous form engaged in gigantic electrical contest. The result is blinding light and flaming heat. The scorching rays from the sun heat the equatorial atmosphere of our earth, where the watery wastes of the ocean are constantly being spun round before the fiery orb of the sun. The rarefied air like a partial vacuum sucks up continuously the water of the ocean and carries it in the form of vapour towards the north and south poles, leaving in its course lavish gifts on all mountain summits and tree-tops. The aqueous vapour condensing to dew and rain collects to form rills and brooks and flows in rivers back again to the sea. But has it not been elevated by the electric force of the sun? Does it not then follow that the active force which drives the mill is of electrical origin? Is it not the electrical sun the mother of the heat and the wind and of the growth of plants and animals and of all terrestrial phenomena?

Let me give another instance:

Wine ferments in the cool cellar and gives off bubbles of carbonic acid.

The mash ferments in the mash-tub and owing to the fermentation, becomes warm.

In both cases electricity is concerned, for the motion of the carbonic acid and the heating of the mash are only an altered form of it, and this is required by rigid logic, when once the correlation of physical forces has been recognized. Still again we ask—how so? And this is the explanation.

Sugar consists of two equivalents of carbonic acid and four of a hydrocarbon, united with two of water, and is in fact a group which has been formed by the action of sunlight, from carbonic acid and water, with the liberation of hydrogen peroxide. It follows that sugar is a compound of "tension-material", forming a whole from the contact of its opposing elements, oxygen and hydrogen. The chemical tension existing in it is a minute fraction of original electricity—of metamorphosed and accumulated sunlight. We thus understand how it is that alcohol and carbonic acid remain united as sugar, only so long as the electric force which holds them together is not lost. But they fall asunder as soon as they are deprived of it by a substance which requires force and material for organic growth, as for example in the case of the yeast cells which require for building their simple forms not only carbonic acid, but also withdraw from the vegetable albumen of the grape-juice, nitrogenous glycocoll. As the yeast cells require force (electricity) for acquiring this new material, they withdraw it from the defenceless grape-juice albumen, which has been set free from the cell walls. The separation thus commenced has an effect similar to that of breaking off a splinter from a pane of glass; the result is a continuous general splitting up. In saccharine grape-juice the electrical transference becomes apparent through the carbonic acid liberated from its electrical tension with alcohol which takes the gaseous form. In other words the fermentation of wine teaches us the highly important fact:

that the removal of carbonic acid and electricity from sugar gives rise to its chemical decomposition, and that when this decomposition commences, that amount of electricity which is not employed in building up a new organism assumes an altered form and manifests itself as heat which can be measured on the thermometer. Thus the decomposition of the sugar gives rise to a liberation of forces, which compels us to regard the undecomposed sugar as a magazine of tensional energy.

This fact which is now universally recognized was, as far as I am aware, first pointed out by the physiologist Hermann. The strange point about the matter is that the chemists and physiologists have hitherto neglected to build further on this fundamental fact.

As it may be proved that the animal kingdom has no other source of energy than the sugar which the rays of the sun have built up in plants, it is conceivable that a rapid decomposition or loss of the substances produced from sugar must give rise to fever, sickness and death.

The occurrence of fever heat is comparable to the production of warmth in the mash tub due to the fermentation of sugar, and this comparison is fully shown by the fact that in fever a much greater amount of urea is found in the secretions, which points plainly to an intense decomposition of albumen.

But in what way does the vital force display itself under normal conditions?

Since we are quite certain that all cases of motion on earth are ultimately due to solar electricity, the effect of which is visible in the water vapour which it raises, as in the descending drops of rain, and which causes the tree to grow and combines the albumen indispensable to the animal creation, it is consequently plain that the decomposition of albumen must again set the solar force at liberty, in a way similar to that in which the falling water turns the mill wheel, as the total of the effective force produced by the water in its descent corresponds according to the law of correlation of physical forces, to the amount of solar heat which was required to elevate the water in the form of vapour.

The same thing is true of these chemical combinations, which in being dissolved liberate physical force. For instance, the carbonic acid condensed in carbonate of lime, when set free by replacement by hydrochloric acid, is when freed from its tension, capable of *raising* the gasholder. It can, however, as in the case of the burning of lime, be as well driven out by heat. And what may be done by heat may also be done by electricity. These forms of energy so imperceptibly pass into one another, that it is impossible to draw a hard and fast line between them.

Continuing the consideration of carbonic acid and regarding the separation of it from chemical combinations as a setting free of a source of energy or motion, we find that the obscurity in which the nervous life of fishes and other aquatic animals has hitherto been involved, is dissipated. For we may observe how sardines, herrings, carps, catfish, sturgeons and the water mammals, the seal and the whale, give proof of a much more intense and enduring power of motion under water, than could be explained as the mechanical equivalent of the small amount of oxygen which is contained in the water. In aquariums it may be seen for what an extraordinary length of time the fish remain under water without rising to the surface in search of air. In place of this, bubbles of carbonic acid rise, nevertheless, from their mouths and gills. And as we may further observe that the numerous aquatic animals mentioned collect considerable accretions of fat, which originate chemically through the splitting off of carbonic acid and water from sugar, we are obliged to acknowledge, that the considerable accumulations of substance which aquatic animals—the oyster not excepted—display, owe their origin to the chemical separation of carbonic acid and water from the nutriment which they absorb.

Similarly to the movements of aquatic animals the exertions of force on our part, are necessarily dependent on a constant splitting up of tissue with a corresponding redintegration so that a regular disintegration of the material at our command, namely albumen seems indispensably necessary to the manifestation of our vital force.

The *orderly* way in which this chemical disintegration takes place is by *respiration*, by means of which oxygen-gas is chemically combined, while other gases are set free in proportion as the chemical combustion proceeds. From this point of view it may be seen that we possess in the albumen of the blood which is composed of adjacent groups of 18 grape-sugar and 18 gelatine-sugar together with earthy substances, a rich magazine of combustible material, and consequently of electrical tension material. Healthy blood albumen represents a reserve capital —a storehouse of current coin.

What a sum of enjoyment may we buy therewith if we proceed wisely—if we act with *economy* and neither allow our treasure to harden behind bolts and bars, nor squander it like the foolish spendthrift and, imagining our wealth to be inexhaustible, set it afire with incendiary torch, or through the hallucination that we can make ourselves poison-proof by the use of poison, become veritable suicides.

We may remain rich, if we make up for the daily consumption which is required to supply our energy, by suitable meals, so as to preserve our solvency. But how utterly poor we may in a short space of time become, if nothing but expenditure takes place without any fresh supply being provided, or when a violent inroad on our resources takes place as in cases of hemorrage, or when total bankruptcy as in the case of attack by cholera, compels us to put all our assets into liquidation and make up the deficit with our lives.

The advantage we derive from our blood albumen, can, through want of knowledge or carelessness be so reduced that we have scarcely any enjoyment therefrom; but at the same time even a *limited* capital can be made to yield so much interest that we fully enjoy our earthly happiness, and may take up the sweet exertion of breathing in both bodily and mental health for a long period of time. The important point is, that the respiration should take place quietly and with regularity, and a sure and regular replacement of the material consumed in respiration will then take place. This regular substitution of material is of course a condition on which the existence of the individual depends, for according to the doctrine of the correlation of forces our nerve substance cannot become active without a loss taking place through the combination of the carbon and hydrogen of our nerve-oil with the oxygen of the respired air to form water and carbonic acid, which pass off as such, diminishing the stock of nerve-oil in the manner of a burning candle. It is consequently plain—first, that the substance used up must be renewed with a certain degree of regularity and, indeed, from the fatty contents of the lymph-vessels, which follow the nerves everywhere as has been anatomically proved.

Secondly, it follows that at the moment when the combustion of the nerve-oil in the oxygen carried to it by the blood is impeded, our life's light must necessarily be extinguished, and that it can only flicker unsteadily whenever the supply of air is inadequate.

The amount of nervous activity finds expression in a definite quantity of products of combustion which form the driving force. In this respect our organisation resembles an apparatus for producing steam for motive power. The same holds true of the brain substance, the oily constituents of which are also liable to consumption. And the offshoots of the brain substance are similarly situated, more especially the pneumogastric nerve which sends out branches to the lungs, stomach, liver, spleen, pancreas and kidneys, so that these organs are in direct connection with the functions of the brain. This is proved by enlargement of the liver after serious vexation, want of appetite due to trouble, grief and care; micturition from fright, anguish and terror; enlargement of the spleen from breathing rarified hot air, deficient in oxygen in the tropics, or through non-absorption of oxygen owing to anaemia and chlorosis. On the other hand we have from imperfect respiration owing to sedentary employment, palpitation of the heart, depression, want of energy and self reliance, nervousness, cowardice, melancholy, spleen, and hypochondria, as the result of an inadequate supply of blood to the liver, stomach, spleen and other abdominal organs.

Ability to work and pleasure in activity are in the main due to the abundant breathing of pure air rich in oxygen. Then the appetite is good and the digestion satisfactory, owing to the secretion of the normal digestive secretions (gall, pancreatic juice, gastric juice, enteric juice). All the glands work properly, and the excretions are normal. Happiness, eloquence, communicativeness, pleasure in art and music, impulse toward action, consideration for our fellow-men are the result.

On the other hand inadequate respiration is the cause of nearly all the ills both bodily and mental to which the flesh is heir.

CONCERNING REST AND SLEEP.

After death when the heart has pulsated for the last time, we find the whole of the blood contained in the body separated into fibrin and serum, and collected in that part of the circulatory system which returns the blood to the heart and which is known as the venous system. The arterial system on the other hand which carries the blood from the heart to all parts of the body is found after death void of blood and filled with air. The opinion having for a long time been held—in fact until the discovery of the circulation of the blood by Harvey—that the same state of things prevailed during life, the arteries which continue the pulsation of the heart and are enabled to do this by their stronger and more elastic coats, received their name of air-tubes, in contradistinction to the blood-tubes or veins, and this nomenclature has survived the error on which it was based. In a certain sense the description of the arteries as air-tubes is perfectly correct, as the vital air or oxygen required for the bodily combustion is carried through the arteries in combination with the blood. By this is not, however, to be understood that the blood of the veins is perfectly free from the vital air or oxygen, but merely that it contains relatively little in comparison with the arterial blood.

The greater part of the vital air which the arterial blood takes up in the lungs and carries to the heart is largely used for burning chemically a certain proportion of the nerve oil and of the muscular tissue and so to produce energy, as already explained. We must here bear in mind that every substance in burning leaves behind it an ash, and we consequently ask "What is the composition of the ash of nerve fat? And what becomes of it?" The chemical composition of this ash is easily enough ascertained. Just as metallic zinc in burning is converted into oxide of zinc or zinc-ash, so the two substances carbon and hydrogen produce two kinds of ash after combustion; the ash produced by carbon is carbonic acid, and the ash of burnt hydrogen is water. Fortunately neither forms of ash are heavy like zinc-ash, but on the contrary inclined to assume the form of vapour. In addition to this they dissolve in the blood, and this enables them to pass through the venous system which runs parallel to the arterial system, first to the right half of the heart, and from there to reach the lungs where they are exhaled in the form of vapour and are replaced by fresh vital air or oxygen.

It is thus clear that whenever respiration ceases to take place abundantly, the ash-products (carbonic acid and water) must collect in the blood; and as all ash substances when strewed over the burning fuel, interfere with the further combustion, so the presence of carbonic acid interferes with the abundant oxidation of nerve-fat. It is to this fact that the feeling of "fatigue" is to be attributed. If for instance the nerve substance of the optic nerve has been kept at work for a length of time say 16 or 18 hours, so much material has been consumed *i. e.* converted into carbonic acid and water that a certain time must be allowed for those products to pass away, whereupon fresh nerve-oil is brought by the lymphatics to these points where room has been made for it. It is the same with the substance of the auditory nerve and the whole brain mass with its processes. All of them are liable to the condition of "fatigue". By putting continuous excessive work upon our organs, even a few hours will be sufficient to fatigue them e. g. the ears. In the case of a strain upon the muscles this takes place even more quickly. Therefore day-labourers who commence work early in the morning do quite right to sleep for an hour in the middle of the day, in order to permit fresh nerve material to flow where it is required, and that the products of combustion may be removed.

In this consumption of material and the necessity for renewing it is to be found *one* of the causes for rest and sleep. The second is the circumstance already mentioned, that our supply of blood is not large enough to fill the whole of the two vascular systems—on the contrary only about half that amount is in circulation. The object of this arrangement is that the blood when flowing forward may find a free course before it, for how could it flow forwards if all the space were occupied? It is, nevertheless, certain, although the statement appears to involve a contradiction to the foregoing, that in all the nerve-regions to which blood flows, the arterial as well as the venous vessels contain blood, for the blood which, for example, flows towards the tips of the fingers must without doubt flow back again through the veins. This apparent contradiction is to be explained by the circumstance that the blood flows towards certain regions of the nervous system in greater quantity at different times of the day, so that a certain sort of ebb and flow takes place, which is facilitated by the way in which the yielding walls of the veins enable them to contain varying amounts of blood. This increase and decrease of the amount of blood is very plainly shown by the spleen which at the time of gastric digestion dilates considerably, but afterwards returns to its original volume. And it is the same with the other abdominal organs which dilate during digestion, as a supply of oxygenated blood becomes a necessity to them to enable them to produce gastric juice, gall, intestinal juice and the pancreatic secretion.

This greater quantity of arterial blood is temporarily withdrawn from other parts of the body, particularly from the brain, so that at a certain time of life, especially in the case also of individuals who are occupied in *mentally* exhausting pursuits during the morning, a certain amount of sleep after the principal meal becomes a necessity.

Unless the brain be plentifully supplied with oxygenated blood, mental productivity is impossible. Power of thought and will diminish in proportion either as the brain becomes overcharged with venous blood, or as arterial blood is kept back from it for the purpose of supplying the organs of digestion. This law is expressed by the proverb *"plenus venter non studet libenter"*.

Similarly the state of sleep in cases of fainting is due to nearly all the blood concentrating itself in the capacious veins of the abdomen, and the brain, in such cases, being subjected to a temporary absence of blood.

Above all, the regular nightly sleep is based upon the fact, that from the time when the optic nerve is no longer excited by the daylight and the auditory nerve by the daily noise, the circulation becomes directed more towards the abdomen, in order to assist the restorative functions of the lymphathics which supply fresh nerve and blood material; for also the nerves which pass to the lymphatic glands and support their functions, are incapable of acting without having oxygenated blood at their command.

The brain gives up a part of the blood that would otherwise flow to it, to the domains of the pneumo-gastric nerve, under the influence of which the lungs, stomach, spleen, liver and kidneys free the blood from the carbonic acid which has collected in it, and at the same time produce fresh lymph, and it receives in exchange this very same fresh lymph that replaces the used up material and renews its capacity for action.

The case of all our voluntary muscles is similar to that of the brain. While not in action their material is not consumed, and fresh lymph flows into them renewing and redintegrating their substance.

The period required for the fluids in the lymphatics to supply blood and nerves with the material required for replacing the waste which they have undergone, is to be estimated at about 6—8 hours. It is for this length of time too that our sleep lasts, and after it is over, the dreamily flickering vital flame shines again with brilliancy, to burn again for 16—18 hours, while thought and clear consciousness re-enter upon their command.

An arrangement of such a kind that the lymphatics suck up the chyle from the alimentary canal merely through the force of capillarity (in which respect they differ from the arterial and venous systems which are acted upon by the pressure of the heart), requires a considerable time—6 to 8 hours—to collect and to fill themselves with the chyle which is conducted from the intestines to a large receptacle—the thoracic duct—by a vast number of fine tubes which unite with one another —as do mountain rivulets to form a brook and brooks to form a river, —an arrangement of this kind makes it plain that the period of sleep can not be shortened arbitrarily without causing the blood material and nerve substance to become exhausted and a corresponding condition of weakness to occur.

The large receptacle into which the fine lymphatics coming from the intestines pour their milk-like contents, is the thoracic duct, which commences at the base of the vertical column, and passes up along it through the thoracic cavity to the level of the collar bone where it empties at the junction of the sub-clavian and left internal jugular veins and periodically mixes its contents with the blood.

The anatomical conditions thus place a certain amount of difficulty in the way of filling the thoracic duct with lymph when the body is in an *upright position*, as the saline lymph which is heavier than water sinks downwards like all liquids in accordance with the law of gravitation; fortunately however the great lymphatic which passes up along the vertebral column is not straight like a gun-barrel but zigzag, which slightly assists the upward movement of the lymph. Still the duct would not fill if we were not in the habit of reclining at night. The thoracic duct then assumes a nearly horizontal position, and it may easily fill with lymph as it does not then require to overcome the pressure of a high hydrostatic column.

During sleep, therefore, it is not merely a question of resting the nerves which during waking hours govern the play of the muscles; by this means alone the organism could not succeed in attaining fresh power. It is of importance that during sleep the body should assume a recumbent position, for when it is upright the lymphatic canal can neither get full nor pour its contents into the blood; and so on awaking there is no feeling of renovation. This supplies an explanation at least in part of the circumstance that continual night-watches, which in spite of over-fatigue send the blood to the brain instead of directing it at the right time to the abdomen, give rise to indigestion, weakness and depression of spirits.

Interrupted sleep is similar in its effects to excessive insomnia. I once knew a man who became very angry if anything occurred to deprive him of his usual afternoon's nap, or if this was interrupted before the usual time by any one urgently desiring to speak to him. On such occasions he would remain in a bad temper for the rest of the day, and his forehead would remain wrinkled till late in the evening although under ordinary circumstances a more light-hearted or better-natured man could not be found. This one example is enough to show to what an extent our state of mind is dependent upon the manner in which the fluids move in the blood vessels and lymphatics. We will proceed to consider this matter more at length.

THE EMOTIONS.

"The King embraced the Minister and gave him the Order of the numsculls with stripes, commanding him to wear the order on his neck and the stripes still higher".

The late Professor of Medicine, Volkmann, published under the name of "Leander" a collection of delightful fairy tales in which a great amount of refreshing entertainment for the spirit is contained. These tales form a veritable banquet for the mind. In one of them occurs the passage above quoted; I will not, however, betray in which one. I will merely state that the quintesseuce of the story goes to show upon what small matters yea, on what utter trifles, our emotional states depend.

The proper nature of the changes in our emotions is without doubt to be traced back to electric currents. Our nerve-fat consists of carbon and hydrogen combined with phosphates. As hydrogen is chemically a metal and any metal will with carbon form a galvanic element or couple which may be excited by a salt, it follows that our vertebral column with its "brain-pole" and its terminal branches which are embedded in muscular or connective tissue, forms a most perfect galvanic apparatus, and in a double sense too. For this apparatus produces independent electricity which can act as an irritant upon and be perceived in other bodies, and consequently resembles an electrical machine. On the other hand it is also sensitive to foreign electricity i. e. electricity coming from without, and so plays the part of an electroscope.

Just as an electric current passing along a wire will give rise to a so called induction current in a wire running parallel to it, so the electric current which traverses our nerves acts upon other existences and exercise a determining effect upon them, in so far as their own supply of electricity is inadequate to enable them to act independently.

Emotions act consequently so to speak like infection. History from the earliest times down to the present day shows this. I may be permitted to insert here two examples as representative of a number of others.

When Aaron was unsuccessful in obtaining the consent of Pharaoh to the departure of the Jewish people from Egypt, and as they themselves displayed no very great inclination to exchange their secure subsistence for an uncertain future, he made use of Fear as an ally in obtaining his end i. e. the making away with the gold and silver ornaments which the Jewesses had borrowed from the good-natured Egyptian women under color of desiring to celebrate their Easter festival in as much splendor as possible. Aaron appeared, suddenly, in the night to the Jews who were assembled to celebrate the Feast, to inform them that Pharaoh had ordered them to depart. (Exodus XII, 31). We cannot say whether he spiced this informations with exaggerations such as that Pharaoh intended to kill all the Jews he found next morning in the country; but I am inclined to believe it. At any rate the passage of Exodus XIV, 5 is in direct contradiction to the alleged command of Pharaoh. It is there stated "When it was told to Pharaoh that the people had fled &c." It may also be concluded from the remorse of the Jews over their departure (Ex. XIV, 11—12) that Aaron had disgracefully lied to them for the purpose of inducing them to take flight. No doubt many of the Jewesses asked "What shall we do with the borrowed jewelry? We must surely give it back first." But it was exactly the gold of the Egyptians that Aaron wanted, he had already determined to melt it up into a lump, at the first opportunity, from which no one would be able to reclaim his own. And he probably replied with affected haste to the honest Jewesses who lived on a friendly footing with their Egyptian mistresses and wished to return what they had borrowed, "There is no time; that can be arranged later. We must be off; whoever remains behind is certain to be killed. Pharaoh will have him speared, broken on the wheel, hanged on the gallows!"

In the middle of the night! Think of the anxiety, the excitement, the anguish and mortal fright which overpowered them. Fear deprives men of coolness and reflection. And if even only a certain number were overcome by fear and wandered about the festive assembly with quaking limbs, that was enough to deprive all the rest of their senses. In this manner Aaron thoroughly attained his object.

As a second example of how fear of disease deprives men of sober deliberation I know of nothing better than the recently established cult (November 1890) of the Koch consumption treatment.

Herr Koch who himself admits that he is largely indebted to his *chemical assistants*, does not possess sufficient acquaintance with organic chemistry to incline me to place reliance in his process. Though it might be possible by means of injecting cyanide of gold, (as with the iodide of mercury which heals syphilis), to interrupt the particular kind of disintegration of tissue to Leucin and Tyrosin, it is nevertheless impossible to cure consumption itself in this way. Experience shows that a diet strengthening and renewing the blood is needed for this. This latter fact is also admitted by Herr Koch, for both he and his followers say "Good diet and treatment are of course also required." But this important and most essential point has but little attention paid to it by those who expect more effect from Koch's nostrum than he himself does. And so sanguine are they that they take no notice of Koch's own statement that "Lupus may, indeed, be cured, but that further experience is required in regard to consumption, and that it must be treated from the commencement* if the remedy is to be of any use. I did not desire to make the matter known as yet, but you are pushing me so much, that ∙I don't know where I stand''. "Let him have a million at once, this benefactor of mankind" exclaim foolish people who have been humbugged by Herr Koch's colleagues, although not a single person has been or *can be* healed by this remedy which works miracles (!) without corresponding generous diet. But what could the unfortunate Koch do?

It is true that the Privy Councillor, Prof. Fräntzel expresses himself quite plainly, for those that know how to read between the lines, to the effect that the case is the same as with the Inoculation for smallpox. The effect will not last for ever, the process must be diligently repeated. His very words are:—"The bacteria decrease in size and vitality after the injection but they still remain capable of life, and we cannot escape the fear, that *if the treatment be stopped too early* even when the patient is apparently free from bacteria, *fresh, fully developed bacteria may appear after the lapse of a few weeks.*"

This is an excellent mode of keeping a way for escape open! If any of the persons subjected to the injection should have a relapse after a few weeks and finally die, we are innocently told "The man did not keep up the treatment long enough, or the injection was not made with sufficient care."

Truly a very convenient means of keeping up the reputation of the discovery! For my part I have no opinion of a business which veils itself in secrecy, and which the persons surrounding the inventor will not make known for fear of losing the money profit they hope to draw from it. One of two motives has hitherto always prevailed in certain callings—either the desire of benefitting humanity or the desire of making money. Where the latter motive prevails I am always distrustful, for the

---

* This limiting condition, "in the first beginning" is characteristic of the whole matter. For what can be regarded as "the first beginning of consumption"?—Shall we consider a cough or a bronchial catarrh, which would terminate of itself in 8 or 14 days, or a hemorrhage from the lungs or stomach as the first beginning of consumption? A man must be quite a genius, who would dare to say: "This condition shows the first beginning of consumption." We may rather say, that the lack of earthy materials (as sulphur, lime and iron) is able to produce in most parts of the body that characteristic disintegration of the connective tissue into Leucin and Tyrocin, which is called consumption; and that we may therefore speak of consumption in all cases, where emaciation shows, that the process of nutrition takes place in an unsatisfactory manner, and so as not to afford a due compensation for the materials consumed. The materials of the body thus gradually *consume* away. The effective cure in such cases consists of a rational mode of nutrition, which continually keeps the end in view, the respiration of cool air, free from dust, suitable clothing and recreations that strengthen the mind. This method will afford a cure, even when consumption has already passed beyond what is generally considered as its beginning.

lust for money hardens the human heart quicker than boiling water does an egg.

It is admitted that Koch's expensive treatment is not designed for the poor classes of invalids. If such poor patients serve as subjects and are cured, it is to the good diet, the secure support and the absence of anxiety for the morrow, as also the relatively pure air of the hospital as compared with the oppressive vapors which so often fill the lungs of poor people with pestilence and predispose them to putrefaction, that this result is to be attributed.

We have all had reason to experience in the cult of Koch what the effects of *infatuation* are. And even members of the medical profession which till now operated with the knife have fallen victims thereto. These doctors now say "Thank God that we have given up the barbarous custom of operating for Lupus with the knife". But it will not be very long before it will be said "It is incomprehensible how such an obscure, untenable method of treatment which besides from the very commencement shunned the light, could set the world in such a state of commotion that without waiting for proof it hailed with frenzied cries of jubilation a variety of barber Jenner's wellknown theory."

For my part I understand it very well. The want of results hitherto attending the ordinary methods of medical treatment had brought the scholastic medicine into great disrepute. It was absolutely necessary to find some means of regaining the prestige which it had lost since Emperor Frederick's death. Then Koch appeared as an angel of salvation—not however for mankind, but for the doctors, for it is they who have raised all the noise which has filled the papers, the newspaper-editors themselves are for once innocent.

Nothing can result of the Koch treatment because it rests upon a false chemical basis, and the bawlers who make such a fuss about it, will produce nothing either*. Florian says quite correctly:

"Les sots savent tous se produire ;
"Le mérite se cache, il faut l'aller trouver."

Trouble and care, as also insufficiency of oxygen in the blood give rise to indigestion, emaciation and finally consumption.

Fear, terror and other forms of mental excitement give rise to diabetes.

Extreme vexation causes swelling of the liver and interruptions of the circulation of the blood in that organ as also in the spleen. Inversely again, defective circulation of the blood in the liver and the spleen give rise to ill-humor and fretfulness. Since liver and spleen are situated in the hypochondriac regions one is quite right in speaking of "hypochondria" as a veritable disease. Another name for it is

---

* This prophecy was fulfilled even before this sheet left the press.

melancholy (black bile) because the secretion of gall in such cases
becomes black in colour. It is true that the bile is secreted by the
liver, the spleen, however, is always affected in conjunction with the
liver, for only the acid venous blood from the spleen passing along the
portal vein to the liver—the structure of which resembles an electric
battery with porous cells—puts it in the condition necessary for carrying
on the process of electrolysis, the most important product of which is
the bile, a substance indispensible for digestion and nutrition. In so
far, therefore, as the spleen participates in the unhealthy condition of
the liver,—spleen, liver. and stomach all being supplied with similar
blood from the three branches of the coeliac artery, it is quite right
to say a man suffers from "spleen", and the real seat of the depression
of spirits is really in the liver, the spleen, the stomach, the pancreas,
the duodenum, and even in the kidneys. The explanation of this is
to be found in the fact that the said organs are connected as by
telegraph wires with the brain by different branches of the pneumo-
gastric nerve.

Hence it comes also, that hunger makes people cross, while it is
generally said that "Well to dine makes a man benign". In this connection too Prof. Volkmann, of whom mention was made at the commencement of this chapter, mentions in his Tale of the Dream-beech,
concerning the fat Landlord of the Crown "He was in a good humor,
for he had just had dinner, and was internally very well lined with
provisions—and that was the most amiable hour of his day."

Yes it is quite true! The best hours come when people are well
sated; they are then happy and light-hearted. So long as men have
enough to eat they do not quarrel nor attempt to upset the established
order of things. And I, therefore, consider that I have done more to
benefit mankind by showing in my book "Das Leben" how all men may
be well fed and thence good, and how sickness may be prevented, than
Herr Koch with his secret poison! I should be very sorry anyways
to stand in his shoes; for after "the Hosanna" of to-day there will
naturally follow very quickly the "Crucify him" of to-morrow.

We will now proceed to speak of still other emotions.

According to the electrical nature of our nerve substance it attracts, holds fast, and assimilates other material which falls within
its sphere of action, and subjects it to itself and regard it as its
own; and so it happens that not only the things which come into
immediate contact with our bodies, but also the objects which daily
fall within our range of vision come to be regarded as in some sort
forming part of the appurtenances of our existence, and we experience
an unpleasant sense of loss when any such object is removed from us.
Even lifeless objects become in time so dear to us that we will not
willingly part from them. How much more then are intimate friends
and among these as the best friends of all, our own family, when we

lose them through death, likely to cause us emotional wounds that in their effects are equivalent to corporal injuries. The sorrow due to such bereavements is not inferior to that caused by loss of property, loss of health, or loss of a limb.

How often does sorrow for a beloved child bring a mother to the grave. How often does one consort follow the other to death since the one cannot find any happiness, "cannot live without the other" as we say.

What we love with devotion and lovingly draw to us, becomes a part of ourselves. It is not an exaggeration to say that when we lose a dear one through death it is as if the heart were torn out of the body. And yet there is a medicine for even this disease, namely work and activity directed to the good of others and designed to bring them happiness.

> „Thou canst live without fellows, nor dost thou need
> The faithful dog or the noble steed,
> Nor the goblets clink, 'mid the jovial throng,
> Nor the merry dance nor the gladsome song,
> Nor woman's lovely face to behold
> Nor wealth nor glory nor glittering gold;
> But all is vanity, grief and strife
> Without love of action, that joy of life!

There are but few who manage to emancipate themselves from the effects of the electric current which, produced by the mass of men, overwhelms and subdues the individual, and they only attain to it in later years. Everything requires time to ripen. Also the knowledge of how little in reality is required for subsistence, health and even happiness, and the knowledge that with small requirements comes freedom and absence from servitude, peace, serenity, happiness and good spirits, all this unless we have special opportunities, is only learned late in life and by degrees.

> C'est à l'ombre de l'indigence que j'ai trouvé la liberté.

In this latter connection it would be a good thing, in order to counteract the ill effects of this universal electric current, by an opposing current, if we could have the maxim which Dr. Koeller put up in his house, framed in every cottage in the land for the purpose of producing independent characters of which the word possesses so few. This maxim is:

> Unto whom a gift thou givest,
> Lord and King thou art;
> Unto whom for gift thou prayest
> Thou a subject art.
>
> But he from whose grace thou seekest
> Neither wealth nor cheer
> Is, though all the earth he ruleth
> But thy like and peer.

This proverb when impressed on our children on starting in life can act as a talisman. A great deal of accumulated electricity is hidden in it, which gives power and strength to the bearer.

To this "talisman" may also be added an "amulet" which is of assistance in every trouble, This amulet too is the property of genial Dr. Koeller's family. Dr. Koeller succeeded in manufacturing turquoises from phosphate of alumina and phosphate of copper under hydraulic pressure. One of his friends took them to Constantinople and they were regarded there by expert jewellers as genuine. These stones are much used among the Orientals as amulets, and they ascribe to them protective power, which as vitreous substances they also doubtless possess, for glass retains electricity and isolates it. These Turkish amulets receive also a *spiritual* protective power by having some motto in terse language inscribed on them. Dr. Koeller's friend had such an adage in Persian inscribed on one of the turquoises and sent this talisman or amulet back to Dr. Koeller. The Persian motto which assists in overcoming all mishaps,—for it is difficult to say, whether the protection is ascribed to the engraved stone, or rather to the inscription which incidentally has a stone for its foundation,—this Persian adage which removes anguish of every kind and bids defiance to every tribulation, is in English:

"Also this will pass by."

NUTRITION.

As the length of the small intestine varies greatly, in the case of adults from 4—8 metres (14—26 feet), the digestive capacity is consequently liable to corresponding modifications. Experience shows that not every form of food agrees with everybody. This manifestly depends upon the chemical composition of the different digestive juices, secreted by the liver, stomach, pancreas, and intestines. The digestion of fatty substances e. g. requires the bile secreted by the liver to be of adequate alkalinity, while the digestion of vegetables like peas and beans whose ash is rich in magnesia requires the action of strongly acid gastric juice or an addition of vinegar to prevent the trouble caused by the formation of insoluble double phosphate of ammonia and magnesia within the intestinal canal. On this account anaemic subjects whose intestinal secretions are always defective are advised to avoid peas and beans as well as acid and fatty foods.

On the other hand foods the ash of which is rich in potash are usually easily digestible by everyone. As examples I may give potatoes and cabbage-turnip the ashes of which vegetables contain more than half their weight of potash. Meat is still more easily digested, its ash being composed of nearly three quarters of phosphates of potash and soda. In pork, potash, soda, and phosphoric acid make up 85 % of the ash, (the remainder being lime and magnesia). This it is which makes the consumption of raw ham so beneficial to anaemic persons.

The common saying that for proper nourishment a certain amount of albumen, fat, and so called carbo-hydrates meaning sugar and starch is essential, depends upon confused views of chemistry. It is certainly true that we have need of fat to renew the nerve-fat which has been consumed, but this can take place without the consumption of fat, for fat can be produced from sugar and starch as well as from albumen. Proof of this is afforded e. g. by oxen which produce more than 100 pounds of fat, while feeding not upon fat but upon starch-containing grasses. The amount of fat laid on by beer drinkers also shows us that malt sugar is a source of fat. In opposition to this, certain beasts of prey would rather starve than take food containing sugar or starch. And there are also men who cannot abide anything sweet. In fine, no universal rule can be established in regard to nutriment; it is much better for every one to follow his natural inclinations (Appetite) as councillor in this matter, and this tells us that we should adopt a certain amount of variety. In the same way as the auditory nerve, and the optic nerve do not continue equally susceptible to the same tone or color, so the intestines also require change if stolidity is not to ensue. By varying the nutriment, effects are produced which commencing in the liver and spleen reach to the brain. The heart too is influenced, beating more quickly, and happiness and pleasure are the result. The pleasures of the table which the members of a family or good friends enjoy together, undoubtedly contribute to health and the prolongation of life; of course they should not be carried to excess. The Persian poet Ibn Jemin says: ..

> "Hearken to the counsel that I give,
> Not in want, nor superfluity,
> But with what is meet, and needful live."

The evil results of overloading the stomach with food affect the health both of the body and of the mind. They are in this respect analogous to the effects of want of food which besides bodily weakness produces ill-humor, manifesting itself now as sorrow and depression then again as anger and rage.

The physiological explanation of this mutual interaction between brain and abdomen is explained anatomically by the fact, that the tenth pair of nerves from the brain which goes by the name of the pneumo-gastric nerve since it is mainly concerned with lungs and stomach, exercises its sway from the head to the abdomen.

In the neighbourhood of the neck its finest branches like cobwebs unite with branches from the seventh pair—the facial nerve—so that the pneumogastric comes to have a connection with the play of the features and with hearing.

Other ramifications runs to the ninth pair—the glosso-pharyngeal nerve which controls the acts of swallowing and tasting.

Others pass to the thoracic portions of the vegetative nerve system —the sympathetic—which gives life to the epithelium of the jaws and causes coughing.

In addition other fibres penetrate the posterior walls of the trachea and the epithelium of the larynx as far as the vocal chords to call forth "the sounds that arise from the throat". (Speaking and singing).

A delicate plexus of the vagus also terminates between œsophagus and trachea near the heart, and participates in the pulsation of that organ.

Other ramifications pass to the branches of the windpipe and the bronchi.

Further a plexus of the two branches of the vagus passes along the posterior and anterior walls of the œsophagus penetrating its epithelium and muscular coat and controlling them. (Swallowing).

As far as the abdomen is concerned a prolongation and division of the plexus of the œsophagus embraces the auterior and posterior walls of the stomach. (Appetite).

In this region the nerve splits up into filaments the fibrils of which are woven in and out like a spider's web and can hardly be detected with the microscope, further they unite with filaments of the sympathetic nerve just as numerous and microscopic, and in this manner obtain a common control over the glands (liver, spleen, pancreas, kidneys &c.) which are actuated by the sympathetic system, as also over the blood vessels densely enveloped by its plexus.

Having thus considered the various regions controlled by the pneumogastric nerve, in consideration of the various openings (mouth, nose, fauces, lungs, windpipe, œsophagus, stomach, brain-cavity, blood-vessels and chambers of the heart) which depend on it for their proper action we must acknowledge the relatiouship between the various processes and activities dependent on the domain of this nerve.

Breathing, coughing, speaking, tasting, swallowing, sobbing, laughing, crying, throttling, swallowing, choking and vomiting, are its work: as also hunger and satiety, want of breath, hoarseness and loss of loss voice, digestion and indigestion, the weak or the strong throbbing of the heart, anguish and pain, shame and blushing, desire and disgust, joy and delight, as well as sorrow and grief are all dependent on it.

These various effects of the pneumo-gastric or vagus nerve, depend largely upon whether it is supplied with a sufficiency of oxygenated blood. No other nervous region is of such importance for our continued existence.

It has been shown experimentally that cutting through even one of the two branches of the vagus will cause in a short time a cessation of circulation in the corresponding lung, giving rise to inflammation of the lung and ultimately death. And when both branches are severed in the region of the neck, the result is immediate death.

Fortunately the opposite is possible up to a certain point. When the appetite is impaired and the enjoyment of life decreased, we need only supply the tenth pair of nerves with oxygenated blood and the whole apparatus will recommence to act properly.

The entry to the region of the vagus nerve which is always open is the lungs. By breathing cool air free of dust, while engaged in a healthy occupation, we may vivify the sympathetic nerve filaments of the arterial system which are intimately connected with the respiratory apparatus. In this manner respiration may produce not only appetite but happiness, enjoyment of life, love of justice, love of our fellow-men, and a variety of other virtues.

On the other hand, respiration of *thin, hot* air causes depression of all the nervous functions producing bad appetite, bad digestion, imperfect formation of fresh lymph and blood, and finally affects the brain, causing a cross state of mind.

A lengthy residence in hot climates predisposes to this condition with almost infallible regularity. Owing to the fact, that movement is there not required to produce bodily warmth, the rapidity of the circulation decreases, and this alone is an important cause of danger. For every impeded movement is converted into heat. And such heat added to the heat of the external atmosphere in the kidneys, spleen, and other viscera tends to cause chemical decomposition of the blood checked in the capillaries and this is in part of an ammoniacal character, causing more or less extended paralysis. It is to this that the deadly climatic diseases are due.

There is no doubt that our best nutriment is the air with its supply of oxygen. We cannot, indeed, eat it, but may be said to drink it. Drinking cool air is equivalent to drinking water, for since the oxygen of the air chemically combines with the hydrogen of the nerve-fat, water is of course produced. In this manner is to be explained the fact that when breathing cool air little thirst is felt, although we experience a desire for fat. In respect to the latter point all travellers agree, that the nearer they approach the North Pole the larger the amount of butter which they are in a position to digest, corresponding to the quantities of fish oil and seal-fat consumed by the natives.

Cool air for respiration is the life giving principle for the soul. Our spirit drinks gas, it is its natural food!

OUR PERPETUUM MOBILE.

That our organism is a very perfectly constructed machine is re-cognizable at the first glance. But every machine has but a limited time capacity for working, corresponding to the amount of fuel with which it is provided. When all the coal or wood has been burned up under the boiler, the supply of steam diminishes and the engine comes to a standstill. Somewhat similar is the state of things with regard to our body. After a certain amount of nerve-fat has been used up a sensation of exhaustion occurs in the muscles and in the brain. We fall asleep, and our bodily machine appears really to stand still. But the next morning it recommences to work, supposing that normal conditions prevail. Must we not then ask ourselves the question how it happens, that our bodily mechanism recommences to work of itself after having stood still for eight hours? This subject has hitherto been a void in the ordinary physiology, which we will now attempt to fill.

The Creator of all things has taken care to provide a special device by means of which a source of energy is produced so to say from nothing, and which bestows upon our organism for a period of from 60—70 years the character of a "perpetuum mobile". This device consists of the spleen, an organ of which even the most highly esteemed physiologists with Dubois-Reymond at their head, declare with emphasis that we know nothing*).

The spleen plays in our system the role of a rejuvenating factor in the sense of a relay-station, and this is effected by means of an apparently insignificant but really important device. The fine terminations of the arteries in all other parts of our body pass into the capillaries of the veins which return the blood to the heart, but in the case of the spleen this is not so. Here the arterial capillaries suddenly cease at a size of $2/10$ mm. (0008 inch) in diameter and turn into small bubbles or sacs called after their discoverer the Malpighian corpuscles. It is plain that under such circumstances the circulation in the arterial capillaries (of the spleen) is brought to a standstill. Since however the ferruginous blood is as discovered by Faraday, magnetic in character, it follows, that the compulsory stoppage of the magnetic blood current must in accordance with the laws

*) A candidate was once asked in a medical examination: "What do you know of the functions of the spleen?" He was much embarrassed and replied that he had known it just before, but that it had passed from his memory. "That is a great pity!" said the examiner, "the only man who has ever known anything about the function of the spleen has forgotten it."

of physics, produce electricity, and proofs exist that this is actually the case.

In the first place the red corpuscles turn into fluid plasma inside the Malpighian corpuscles, an action which is analogous to that which takes place when electricity is passed through blood which has been taken from a vein. From this circumstances we may conclude by analogy, that the impact of the magnetic blood current upon the walls of the Malpighian corpuscles has the effect of converting the magnetism into electricity. We are therefore obliged to conceive that minute electric discharges take place from the spherical walls of the Malpighian corpuscles into the blood. This source of energy, "Electricity", in so far as it is due to the coming to rest of the magnetism i. e. of the circulation of the blood may be regarded as originating out of nothing. That accumulated electricity does display itself in the manner described, is shown secondly, from the fact that the fluid plasma which penetrates through the delicate membrane of the corpuscles contains as products fatty acids (formic, acetic, propionic, and butyric acids) which are the results of electric decomposition. These fatty acids have been produced by the chemical decomposition of nerve-fat. In fact the delicate terminations of the nerves of the spleen which proceed from the solar plexus of the sympathetic system, accompany the ramifications of the splenic artery step by step even to the Malpighian corpuscles. Where the arterial capillaries stop, the nerves stop also, and recommence with the venous capillaries running parallel to them. Now, as in the case of the electric incandescent lamp small incandescent particles of carbon are projected from one pole to the other, so the electrolytic decomposition products (in the form of formic acid, acetic acid with other fatty acids) pass into the rejuvenated plasma (pulpa splenica) which is thereby rendered acid. The venous capillaries are supplied with this acid splenic secretion, and their contents are of course acid, and it is finally carried through the portal vein to the liver. Now as all acids act as electrical excitants, this blood from the portal vein must supply the necessary electrical tension to the hepatic cells (*acini*) of the liver so as to secrete a strongly digestive bile, which owing to its alkaline nature is the electro-chemical counterpart to the acid splenic blood. The action of the liver is thus also dependent on that of the spleen. Without the spleen no bile secretion; without the secretion of bile no absorption of chyle from the chyme, and consequently no proper nourishment of the system, a diminished new formation of blood and nerve substance, and thence a successively increasing weariness.

Since now the absorption of chyle takes place principally at night at which time our bodily machine *appears* to stand still, it follows that this stoppage is only apparent; the truth is that the machine is working in the region of the abdomen to produce fresh fuel.

When in hot climates the inadequate respiration of oxygen produces instead of a swift exudation of the plasma through the Malpighian corpuscles, a stoppage of circulation and swelling of the spleen, the natural relay-station which under ordinary circumstances supplies the splenic blood with electrical energy in the form of formic and acetic acid is thrown out of order. Under these circumstances our bodily machine gradually ceases to be a "perpetuum mobile". The original disorder of the spleen results in liver-complaint, indigestion, constipation and all the numerous disturbances known under the heading of climatic fever.*

* On the day when the proof of this sheet was being read (Febry. 27 th 1891) the newspapers brought the following notice:

"Sad news come from Kamerun. The official physician Dr. *Hugo Zahl*, born in Bromberg died Febry 12 th of climatic fever. Death ensued, on the journey to Lagos on the steamer "Adolf Woermann". Mr Zahl, who had been active in the imperial service in Kamerun (Cameroon) for the last three years, only reached an age of 35 years."

In such a way the immortal *"ex cathedra"* declaration of the physiologist Dubois-Reymond is avenged: "Gentlemen, we now proceed to the *spleen*. We know nothing of the function of the spleen; we pass on to the *Liver*."

The ignorance of physicians hitherto of the fact that the spleen acts as a source of electricity has prevented so far any inquiry, how the spleen when it refuses to act may through rational means be restored to its activity. I myself have solved the question both theoretically and practically, and this has also brought its blessing in many places, but the matter has not yet become *generally* known. On this account I shall in the following give particular attention to the consideration of *climatic affections*, as well as to *consumption*, especially since the climatic fevers themselves cover almost the whole scale of pathological states.

III.

PATHOLOGY

DISTURBANCES OF HEALTH.

A. IN GENERAL.

The want of clearness which prevails in the department of physiology where physical, chemical, and electrical "stimuli" of the nerve-apparatus are frequently spoken of, is of course again to be met with in pathology. By a chemical stimulus is meant for example, touching a muscle with a solution of salt. Since however salt acts as an electrical excitant, electrical and chemical stimuli come to the same thing. There are also other considerations which lead us to the same result. For as far as electricity is concerned, it only *appears* to come out of the conductory wire; as a matter of fact the wire only transports the electricity which in the first case was produced by the interaction of opposing substances either acids or salts or various metals. And in so far as the electricity is merely a species of motion, it follows that the physical "stimuli" (pressure, blows, touch &c.) are in reality of the same nature as the electrical. We must remember that this sort of artificial subdivision has lost its significance since the identity of all forms of energy has been established.

The artificial subdivision due to one-sided views must also yield in the department of pathology. I am of opinion that we can put the matter very shortly and say "Health is the continuance of the cohesive electric force, while illness is the diminution of it, so that restoration to health is the same thing as a recuperation of electrical energy. It is only in accordance with this standard, that therapeutics or healing art can apply the right remedy.

We will therefore keep to this principle;—Health depends upon the keeping together of our bodily substance owing to the continuous action of electrical energy, while disease is characterized more or less by a chemical disintegration or disruption of the same, as a preliminary to death.

To make the matter plainer, I will draw attention to the decomposition of dead bodies which is characterized by a disagreeable odor. This bad odor is due to the fat which composes so large a portion of our bodies being no longer held in combination with carbonate of ammonia and the earthy bases, since the electrical energy which caused their orderly combination into an organism has disappeared. In consequence of this fact considerable quantities of gas or spirits,—ammonia, carbonic acid, ammonium sulphide, butyrate and capronate—are set free. These gases are the principal cause of cadaverous odors. Besides them, however, we have the decomposition products of sugar and gelatine i. e. lactic acid, leucin, and tyrosin. Similar products are produced when meat is digested in the stomach—digestion being a species of decomposition; and this explains the offensive odor attaching to the contents of the stomach when ejected by vomiting. On the other hand such decomposition products are not produced from the muscular and connective tissue of an intensively electrical living body; they arise only in cases of disease, and the substance from which they are produced is essentially gelatine which consists of fat 2 (C_6 H_{12}) and dehydrated ammonium carbonate ($2CO_2N_2H_4$). This gelatine is distributed throughout the whole of the body though in one place more than in another. For instance, gelatine can be extracted from the bones, the connective tissue, the sinews and from the intestines.

The nerve-fibres contain relatively little gelatine, being composed almost exclusively of fat together with ammonium phosphate. Since now the ammonia and the phosphoric acid can with oxidized fatty substances produce gaseous bodies e. g. ammonium acetate and phosphoretted hydrogen, which latter gives rise to part of the evil odor before mentioned, it follows that the nerve substance in spite of the protection against a too rapid consumption with which it is provided, is still liable to decomposition since it has as chemical basis the general material of the body namely fat and gelatine. Through these two substances in combination with phosphoric acid, without which neither growth nor conservation of the organism are possible, bones, muscles and nerves appear also pathologically to be related. The chemical distinction between them is essentially due to the fact that in bone phosphate of lime, in muscle phosphate of potash, and in the nerves phosphate of ammonia are glued to combustible substances partly carburetted hydrogen and partly cyanogen by means of gelatine.

This similarity of chemical composition explains how it is that when any particular region falls a prey to chemical decomposition, others are quickly similarly affected. Chemically considered the body is a whole, and it is only from the point of view of anatomy that it appears to be composed of parts. The characteristic chemical basis is gelatine by means of which our bones, muscles, and nerves are glued together internally and to one another. This fact is of as much importance for pathology as for physiology. The conceptions of gelatine or glue, and of clay occur side by side in the oldest writers. That from the coagulated casein of milk "casein-gelatine" is prepared and that the child obtains his "blood-gelatine" from the casein of milk is expressed in Job X, v. 9 and 10.

"Remember, I beseech thee, that Thou hast made me as the clay, and wilt Thou bring me into dust again?

Hast Thou not poured me out as milk, and curdled me like cheese?"

The basis (gelatine) of our bodily substance is, as growth goes on, formed from the gluten of the grain, from milk, then from the adhesive egg-albumen, from all kinds of vegetable albumen, and from the gelatine of bones in the form of soup. When this gelatine is subjected to chemical decomposition without adequate replacement, the bodily elasticity due to the gelatine is lost, and with it the most important source of energy.

As this fact has not hitherto, as far as I know, obtained adequate recognition, and as my system of pathology and therapeutics is based upon it, I think it advisable to draw attention to the following facts.

The gluey jelly obtained by boiling calves feet in water and cooling the decoction, vibrates elastically at the slightest touch. This elasticity is the characteristic peculiarity of gelatine. It constitutes a continuous automatic source of energy which costs, so to speak, nothing as nothing is used up from it. The proof of this is afforded by gelatine in its most concentrated form of whalebone. But even when combined with water in the form of cartilage and tendons, we have numerous proofs that this fundamental gelatine forms an inexpensive source of energy. If for instance we bend the cartilage of our nose or ear out of its natural position, the organs in question return to their original place without any effort on our part; for in this consists the conception of elasticity, the possession of which quality of elasticity on the part of gelatine, is without doubt due to its ammoniacal basis. As further examples of this automatic elasticity of gelatine I will mention the sinews of the race-horse, the hare, and the kangaroo; further the upright position of man attained without trouble, which is due to the elastic nature of the different bands surrounding the separate vertebrae, as may be seen in carving the back of a hare.

This spring-like character of gelatine is made use of in various organs of the body as an inexpensive source of energy e. g. in the lungs. The amount of cartilaginous substance which they contain, is noticed by house-wives in preparing a hash of lights (lungs). And it is to this elastic gelatine that we owe the equable dilations and contractions of the lungs during inspiration and expiration; and the results due to diminution of gelatine in the sinews and its effects on the automatic movements of the lungs may be easily imagined.

The amount of gelatine which must be cut out of kidneys when they are prepared for the table, is also well known to housewives. Almost richer yet in tendinous matter is the spleen, which is quite interpenetrated by it. And the intestines when dried consist of little else. From the intestines of hens and cats gelatine may not only be boiled out, but they may be manufactured into all kinds of fiddle and bow strings. The bladder is similar in this respect to the intestines; it consists essentially of tendinous matter. Thus the network of gelatinous tendinous matter extends throughout the whole body from the brain to the intestines. There is no region destitute of a certain amount of gelatine, and as our body consists of three parts water, it may occur that the gelatine may begin to putrify or decay even during life, owing to the fact that the parts affected begin to disintegrate chemically, as may be observed in scrophula, syphilis, small-pox, anthrax and plague. The ground of this chemical disintegration is in every case and especially in the disintegration of the gelatine of the blood, a diminution of electrical energy. We must if we wish to retain or regain health, take care to bring into action those factors which produce electrical energy. The principal element for this is the oxygen of the air we breathe. Through its chemical combination with the ferruginous gelatine of the blood, electricity is produced, as in all cases of chemical combination. The electricity so produced tends to flow to a distance. It accordingly proceeds along the spiral filaments of the sympathetic nerve, which helps to form the walls of the arteries right on to the capillaries which surround the terminations of the nerves; and since at the same time an induction current is set up in the nerve strands which run parallel with the blood-vessels and this also tends to diffuse itself, it follows that the electricity so produced proceeds to the ultimate ramifications of the nerves where it manifests itself in the form of attraction. The contents of the blood vessels are affected by this attractive force, and its ferruginous hemoglobin being surrounded by the spirals of the sympathetic nerve becomes magnetized and consequently attracted and set in motion by the spirals of the sympathetic nerve as well as by the terminations of the cerebro-spinal systems. All this is due to inhaled oxygen. Should this most important source of electricity (respiration of oxygen) be impeded or diminished by any cause, the amount of electricity produced and consequently our vital force, becomes diminished in the same proportion. It then happens that the process of chemical decomposition which ought only to begin after death, even during life attacks certain organs, and most quickly those which require a considerable supply of air, namely the lungs and the dependent respiratory apparatus. Thereupon the tissue of the lungs decomposes in spots, as in ordinary dissolution into Leucin and Tyrocin, which owing to the whitish appearance produced, resembling cheese, is described as "cheesy degeneration", and since this degeneration goes on spherically from certain centres, the name of tubercles has been given to the spherical lumps, and tuberculosis to the disease. It is in fact want of chemical knowledge which centuries ago gave rise to this name, which is still adhered to.

Simultaneously with these visible knots or lumps, other commencements of chemical decomposition are observable which however escape the naked eye; they can be seen only with a microscope and are discoverable in the first place as crystalline rods so delicate, that they must be dyed with aniline before they can be plainly distinguished from the surrounding material. It is plain, that a limit can be set to this commencement of the decomposition of blood, through breathing good air, and bringing the blood back again to its normal condition by suitable nutrition. In this case the fragments of leucin and tyrocin are converted by the lymphatics into proper constituents of the body, and in this way the beginnings of consumption, or rather what if neglected would in time become consumption may, be cured.

Herr Koch, however, is of quite a different opinion on this point. He like all so-called "bacteriologists" holds the extraordinary view, that the microscopic crystalline fragments of decomposed blood albumen are mysterious entities—half animal, half plant—which in troops invade the organism from without, as the Huns and Vandals once upon a time overran foreign territories.

In reality the minute crystals described by Herr Koch and his followers as "bacilli" or "bacteria" (meaning little rods), are to be regarded as products of incomplete oxidation of the blood due to insufficient respiration. This process corresponds to the production of fatty acids from neutral tallow-fat when the latter is slowly and imperfectly acted on by oxygen. When the supply of air is adequate, the tallow is rapidly, completely and without odor oxidized into water and carbonic acid, but where the supply of air is weak and impeded, various decomposition products of fat, such as in part are also found in rancid butter, are produced, for example capric acid $C_{10}H_{20}O_2$, caprilic acid $C_8H_{16}O_2$, caproic acid $C_6H_{12}O_2$, butyric acid $C_4H_8O_2$, propionic acid $C_3H_6O_2$, acetic acid $C_2H_4O_2$, formic acid CH_2O_2; but more complicated products also result namely palmitic acid $C_{16}H_{32}O_2$, stearic acid $C_{18}H_{36}O_2$, arachidic acid $C_{20}H_{40}O_2$ which chemically detached remnants also occur in rancid butter, all of them the results of incompletely oxidized animal fat ($C_{42}H_{84}O_8$).

In speaking here of "splitting up" I employ the expression literally in the sense in which chips are split off from a block of wood. Every group composed of a larger number of atoms is capable of being thus split up. One of the best known instances of this is the splitting up of the molecule of sugar $C_6H_{12}O_6$ into two equivalents of lactic acid $C_3H_6O_3$ in consequence of an inadequate supply of air. This cause—inadequate oxidation in consequence of an insufficient supply of air is well known in technical industries and is utilized for producing decompositions.

I will only allude to the occurrence of lactic acid in the liquor from the mash-tub in beer brewing, which contains malt-sugar, and with which the air was prevented by the water vapour which was given off, from properly coming in contact, in the primitive method of cooling employed. In the same category is the occurrence of lactic acid on the cut surfaces of the beet and of a peeled apple.

As now in the fermentation of fruit-sugar on the one hand lactic acid COO, CHH, CHH, HHO, on the other malic acid COO, CHH, CHH, HHO, COO or succinic acid COO, CHH, CHH, COOHH, or glycerine HHO, CHH, CHH, COOHH, alcohol HHO, CHH, CHH, and carbonic acid COO may be produced, so similarly chemical decomposition products, with a varying number of atoms and various characteristics, may be produced by the decomposition of the albumen of the muscles and the blood. The composition of these products depends to a certain extent on the nature of the glandular tissues with which the gelatine was chemically combined.

In my book "Das Leben" I have explained in detail, that the chemical basis of our bodily substance rests upon altered groups either of cyanogen CN or hydrocyanic acid HCN. I give illustrations of these groups in reference to the bases creatin, sarkin, xanthin and urate of ammonia.

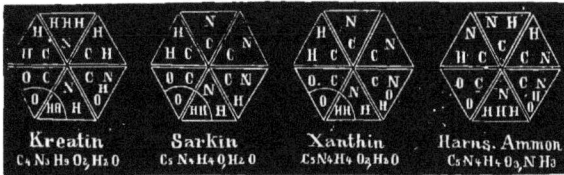

Kreatin	Sarkin	Xanthin	Harns. Ammon.
C₄ N₃ H₅ O₂ H₂ O	C₅ N₄ H₄ O H₂ O	C₅N₄H₄ O₂H₂O	C₅ N₄ H₄ O₃ N H₃

In this way it is most easily explained how, like to the production of lactic acid from grape-sugar, the gelatinous tissues of our body may give rise to substances that in part bear the character of ammonia, in part that of hydrocyanic acid. Doubtless these decomposition products are of poisonous nature in their origin and only become innocuous through chemical combination with water. In the same way the poison, which Herr Koch prepares by causing gelatine mixed with extract of horse flesh to decompose in chemically the same way, on dilution with water soon loses its poisonous properties. This latter reaction corresponds to the combination of poisonous ethyl cyanide to form a fatty ammonia compound namely propionate of ammonia.

HCN, CHH, CHH + 2HHO = CHH, CHH, CHHOO, HHHN

Similarly hydrocyanic acid with two molecules of water forms formiate of ammonia COOHH, HHHN.

Conversely a poisonous substance of the nature of prussic acid may be produced in our organisms by the withdrawal of the constituents of water from the partially oxidized hydrocarbons of gelatine and the ammonia. Such a reaction becomes comprehensible from the consideration of the method by which I have artificially produced coniin C_8 H_{15} N. This anhydrous substance is obtained when absolute alcohol $C_2 H_4$, $H_2 O$ acetone $C_3 H_4$, $H_2 O$ and anhydrous ammonia are allowed to react upon one another, the ammonia combining with one molecule of alcohol and 2 molecules of acetone, and this is accompanied with the detachment of 3 molecules of water.

Poisonous as coniin is, it can be turned into innocuous butyric acid by the action of water and air, when the ammonia has been withdrawn by diluted acids. Three hydro-carbon groups then unite, to form the common group $C_8 H_{12}$ similarly to the production of fusel oil $C_5 H_{10} H_2 O$ from aqueous solutions of tannin in the presence of calcium salts. The new group $C_8 H_{12}$, in the presence of alkali and air takes up two proportions of water, whereupon two proportions of butyric acid $C_4 H_8 O_2$ are separated. In this case acids, alkalis, water and air act together to deprive the coniin of its poisonous properties which resemble prussic acid. Acids, bases, water and air—these remedies must be borne in mind in all therapeutics as soon as we have learned:

1) That from innocuous fatty acid combinations with ammonia poisonous substances of the nature of prussic acid may be produced by dehydration.

2) That inversely prussic acid compounds may by the action of water together with acids or bases be reconverted into innocuous fatty combinations.

3) That the incomplete oxidation of sugar, fat, gelatine and albumen produces a chemical separation of their atomic groups.

Insofar as the said substances (sugar, fat, albumen,) form a connected trine, as one of them can be produced from the other, it becomes explicable from their inner mutual relationship that both separately and together they are' subject to the common fate of chemical decomposition.

I am convinced that if Herr Koch were acquainted with these chemical facts, namely that from an ammonia compound and consequently from our blood through the effect of electro-chemical reactions, prussic acid and water may be produced, as is actually the case in the "cyanosis" and "hydropsy" the "siamese twins" of pathology, he would cease to regard the bacilli, that as decomposition products of blood have the character of poisonous cyanogen compounds, as the causes of diseases, but rather as the results of disease, and would then abandon his extraordinary system of pathology which consists of employing the poisonous decomposition products of blood-albumen—the bacilli—as a remedy for tuberculosis i. e. to cure the continued decomposition of blood-albumen. On the contrary he would follow the natural course and seek by the application of oxygen to stop the incomplete oxidation of the blood and the production of the poisonous substances which result therefrom. He would also seek to cause the bacilli to take up water and so lose their poisonous prussic acid character, and become neutralized and innocuous by the action of the natural antidote to prussic acid namely ferrous oxide. Finally he would take care by improving the constitution of the blood by administering proper nourishment containing proper basic substances (iron, manganese, lime, magnesia, soda) and salts capable of combining with albumen, that the disease should not easily recur i. e. that the patient should not have a relapse.

Hitherto Herr Koch appears still to have regarded the bacilli as living creatures that are nourished like other living creatures and give off what they do not digest. In this sense Herr Koch speaks of a change of substance in the bacilli, although they have neither intestines, mouth, anus, or urethra or any other organ, for they are nothing but the finest crystalline needles. It is consequently very rash to speak of a change of substance on their part. The same is true of the immigration of numberless bacilli. The truth of the matter is that in the same way as a rotten apple i. e. an apple which is undergoing the lactic acid fermentation will "infect" others, so the process of fermentation or disintegration which produces the prussic acid bacilli may be communicated to healthy tissue where its power of resistance is not sufficient, as is the case in a small guinea-pig. The chemical process in cases of decomposition which has been started by infection, is like vinous or lactic fermentation which after having been once commenced, continues as long as decomposable material is present. Consequently it is not permissible to hold or to propagate the view that the bacilli increase by way of reproduction like as a cow brings forth a calf. To suppose this is sheer nonsense as the bacilli are just as little provided with organs of generation as they are with stomach or intestines. The rapid way in which they increase in the albumen of the blood while in circulation and in the tissues, is to be explained by the impulse towards electrolytic decomposition which has been given in a similar way to that in which a single piece of glowing coal suffices to occasion the chemical decomposition of a whole barrel of gunpowder into sulphurous acid, carbonic acid, cyanogen and nitrogen.

Now in the same way in which a special form of gland determines the decomposition product into which the blood shall be converted e. g. as another secretion is produced from the salivary glands of the mouth than from the pancreas, so we find in the lungs different decomposition products from those which are formed in the spleen, in the thymus gland and in the intestinal glands.

It is this circumstance which enables a poisonous decomposition product due to the degeneration of substance, say in the intestinal glands of one individual, to give rise to similar decomposition in the corresponding organs of another individual, and no doubt in the same way lung-tuberculosis may be communicated to individuals whose capacity for resistance is not very great. It is on this ground that Herr Koch has been induced to say "I inject bacilli of tuberculosis into a guinea-pig and the guinea-pig thereupon gets tuberculosis, *therefore* the bacilli are the cause of tuberculosis. However correct this conclusion may appear at first sight, it is nevertheless false because *one-sided*. Tuberculosis also arises *without communication from without by internal chemical degeneration* due to want of electricity, water, oxygen, and earthy bases, and *scarcely any other causes* than these are active when it is a question of tuberculosis in man and animals.

The conception of "infection" will continue to suffer from a want of clearness so long as no explanation is given of it. In this respect I may point out how easily we may make mistakes if we only admit the possibility of *one* explanation. We will suppose that a fire breaks out at a manufactory which ends by laying it in ashes, and that after the breaking out of the fire two boys were seen running off. Would one not be inclined to suspect the boys of incendiarism? And yet the fire might have originated in a totally different way, for instance according to the following receipt:

Take some woolen rags that have been used for months to clean machinery with oil; throw them all on a heap and let them remain neglected close to some wooden wall. You can then be quite certain that a fire will break out sooner or later, since the oil in the rags undergoes chemical changes which produce heat and end in combustion.

One might even give the exact amounts required for such a receipt, for instance:

"Take 20 lbs. of sheeps wool, 1 lb. of tallow and 6 lbs. of linseed oil. Pound the whole well together, tie up the whole tightly with strings and leave it. Return to it after six days. If by that time the bundle has not been reduced to charcoal, it will certainly take fire as soon as you open the door and admit air to it."

I wish in this way to show that communication from without is not the only method of starting the process of combustion or of decomposition, but is on the contrary exceptional. Much more frequently —in fact almost always, internal decomposition is the cause of so-called infectious diseases.

One need only enter certain cow-houses, from which one almost falls back in a swoon, to recognize the cause of disease amongst animals. *Almost all the oxygen has been consumed by the respiration of the animals.* Is any further explanation required, why so many cattle are affected by tuberculosis? What would be the good of inoculating such cows with Koch's lymph? It would be much better to send them to the camp hospitals in Moabit where they would breathe fresh air and get healthy provender. Under such circumstances they would recover *even if* injected with the lymph, since large atomic groups will not perish from a small quantity of prussic acid poison.

Only one of the two things can be true—*either* it is Koch's injection which cures the tuberculosis, and then the hospitals are superfluous, *or* it is the oxygen of the air and suitable nourishment which effect the cure and then the injection may be dispensed with.

What will the history of medicine say some day to this episode of "Koch"? I am of opinion that we may learn from it that during life, honor and reputation stand in inverse relationship to real merit, but also, that false fame like the parabolic orbit of the comet rapidly loses in brightness, whereas true glory increases in proportion to the square of the time that has elapsed.

How astonishingly little is required to enable a single man, who is credited without criticism or examination, to inoculate the whole globe with a false belief which may last a thousand years and a half thousand more, is proved from history by the Ptolemaic doctrine of the motion of the heavenly bodies. According to this theory the sun races around the earth through the mundane space every 24 hours, which would give it a rate of 100000 geographical miles a minute. This opinion was held for nearly 1500 years until Kepler discovered the truth. But for discovering the truth he was not only not paid a million, but his professor's salary even was withheld from him, so that he found himself obliged to put a copy of his work "On the planet Mars" in his pocket and trudge on foot to Regensburg to plead his case before the Reichstag. This occurred in the middle of November. Drenched by rain, chilled by the November storms, tortured by hunger, chattering with fever, and a prey to grief,—was it any wonder that Kepler under such circumstances should have taken tuberculosis? On the 15 of November 1630 he fell a victim to death in the Imperial City of Regensburg. But restitution has come at last! As a compensation for the indolence of the world in respect to the man who had discovered the movement of the heavenly bodies, 360 years later an imitator of the English barber Jenner is overwhelmed with honors by the German Reichstag in spite of the fact that he is neither astronomer, mathematician, nor chemist, because the idea has gained ground that he can cure tuberculosis, from which disease Kepler, in my opinion, died.

That the Minister of State von Gossler made matters smooth for Herr Koch does not confer any special value on his discovery. Herr von Gossler was in the same boat with the numbers of the doctors who came to Berlin hoping to learn how to cure consumption. But ought not Herr Koch, ignorant as he is of the chemistry of the decomposition of albumen to have been afraid to act as a guide to others?

That the principal source of the electricity, which holds our bodily substance together, is due to the oxygen in respiration combining with certain constituents of the blood albumen, to produce in accordance with the laws of chemical combination a proximately constant amount of electricity—this we saw proved by the fact that on the cessation of respiration the bodily substance is subjected to chemical decomposition producing disagreeable odors.

Other factors also conspire with the respiration of oxygen to produce electricity. As such a factor we must regard the saltiness of the blood.

It is well known that butchers' meat can be preserved from putrefaction by salting it, and that common salt was originally used to embalm the bodies of the French kings.

This faculty the salt possesses of protecting flesh from decomposition is also to be attributed to electricity.

A copper and a zinc plate develop electricity when brought into contact with blotting paper which has been soaked in a solution of salt. Similarly our bodily substance also produces a certain amount of electricity through the presence of salts and moisture because of its contents of cyanogen or hydrocarbon compounds, corresponding to a galvanic pile of carbon and hydrogen plus carbon and nitrogen. To these salts contained in the blood, in addition to common salt which makes up the half, belong phosphates and sulphates of potash and soda, while the electricity of the nerves is produced by phosphate of ammonia.

As a third element in addition to the respiration of oxygen and the salts which give electricity, the earthy constituents of our bodies producing electric tension (potash, magnesia, manganese, and iron) are to be considered.

The importance of these earthy bodies has been alluded to at the commencement (pp 3 to 5). And this is the place to point out characteristic facts facilitating the comprehension of the universally ruling laws of sustenance.

According to a traveler who described to me his experiences on the coast of Brazil, the females of that country even at an age of 10 years lose all their teeth, so that a European woman who lands there with a full complement of teeth is regarded as an extraordinary phenomenon. The explanation of the early loss of the teeth is to be found in the fact that the inhabitants of Brazil mainly live on bananas and other sweet fruits which contain only small traces of phosphate of lime, and it is this with gelatine that forms the basis of the teeth. The Brazilians, indeed, also eat meat, but meat is also poor in lime, containing only so much as corresponds to the blood-vessels penetrating the muscles and the red corpuscles in it which are based on sulphur, iron and lime. Thus it is that meat after burning leaves an ash containing about eight times as much potash and soda as lime and magnesia. Potash and soda have the faculty of removing lime, and magnesia and the oxides of manganese and iron from their sphere of action; they consequently also precipitate them from solution. It is due to this fact to a certain extent, that phosphate of lime is separated from the tissues, and aggregated in the bones, which unlike the tiger and the dog we do not eat. Of the tiger it is said, in the childrens' Reader, "he devours flesh and bone together". If he did not do that he would suffer from asthma, corpulency and rheumatism, like the pugs who are fed by their mistresses on meat without bone, for without lime no red blood corpuscles can be produced; and red corpuscles are essential to proper respiration, without which asthma occurs, since it is their function to chemically combine with the oxygen; and without proper respiration the fat is not consumed, but aggregates in proportion as it is produced from sugar and albumen by the chemical separation of carbonic acid. Hence corpulency. Finally the blood pressure decreases when the supply of oxygen is inadequate, and the circulation tends to come to a standstill now in these vessels now in others, and with the interruption of the circulation of the magnetic blood, electric excitation of the nerve fibres occurs, which is rheumatism. Thus it happens that people, who prefer a meat diet together with plentiful indulgence in beer and wine, in course of time, when the proportion of sulphur and lime in the blood has fallen below a certain measure, fall a prey to asthma, corpulency, and rheumatism.

Without lime, firm bones cannot be formed. Blood vessels enter every bone; they nourish it and effect its change of substance. In this way they supply it with the necessary phosphate of lime which is to a certain amount soluble in the salty serum of the blood. A blood vessel runs into every tooth as into every bone, to supply it with nutriment, while a vein returns from it. If now the blood has not enough corpuscles containing lime in it, the artery is unable to supply lime to the tooth; on the contrary, the vein constantly carries some of the phosphate of lime away with it, and in proportion as this loss of lime goes on, the gelatine (cement) loses its chemical support, falling a prey to decomposition and decay. In this way the teeth are first of all attacked from the inside by the vein, thus becoming hollow while the saliva attacks them from without, dissolving the softer basis of the teeth—the gelatine—so that robbed of their support in both directions, they ultimately break off.

The lime represents a certain quantity of electrical cohesive force, forming a safeguard against the chemical decomposition of the gelatine, and this electrical tensional force is exercised to hold together 18 hydrocarbons, which in the oxidized state as stearic acid $C_{18}H_{36}O_2$ form with lime a hard soap insoluble in water. The zone of activity of the single lime molecule Ca_2O does not extend beyond the 18 hydrocarbons. Accordingly it is easy to understand that, when a certain definite amount of lime exercises only a certain definite amount of force, the strength of our sinews and muscles attached to the bones, must vary according to the strength of our bony structure. Firm, strong flesh depends on firm and strong bones. Thus the lime proves itself to be in reality one of the bases, on which the holding together of the human body depends. If we are to retain health and strength we must have food which contains lime. It is true that some people maintain that food and water containing lime tend to ossify the arteries, but this is a very one-sided and altogether incorrect view, for ossification of the arteries only occurs in those cases in which the blood does not contain sufficient salt to keep the phosphate of lime in solution, for instance in the case of vegetarians who eat too little salt, and also in the case of systematic wine drinkers. Wine contains no salt—on the contrary it diminishes the indispensable salt, owing to the increased amount of urination which it causes. This must naturally cause the phosphate of lime deprived of the salt which keeps it in solution, to crystallize out in the smaller arteries. Where this ossification of the arteries is observable, the best remedy is a mineral water rich in salt.

Since the sweet fruits of the tropics contain but few salty particles, they can have but little phosphate of lime in them. It is otherwise with grasses. They extract considerable amounts of salt from the soil which through its moisture stands related to the sea-water, and they consequently contain phosphate of lime. Hence the buffalos which live on the prairie grass, have a massive bone structure.

Where lime is absent, only boneless creatures such as worms and maggots can exist. In the latter connection it is of importance to notice the relation between certain phenomena, as we have already done in the case of asthma, corpulency and rheumatism.

That the women of Brazil in consequence of the absence of lime in their diet lose their teeth early in life, is not the only disadvantage which arises therefrom. It is also closely connected with an increase in infant mortality. The traveller, whom I already mentioned, told me that his wife visited a neighbour to inquire after her health, and that of her child which was born some 10 days before. She was informed that the child was not in good health, and that it constantly discharged innumerable worms from the intestines. Whence came these worms? The child had had nothing but the mother's milk, and it is just there, that we must seek for the cause. Since the blood of the Brazilians is wanting in earthy constituents, the phosphatic epithelium of the intestines of the child became converted into worms. That even in our climate the dilution of cows' milk with water, diminishing as it does the percentage of salt and lime, gives rise to cramp and worms in children has been already pointed out on page 5. In such cases the cramps are not caused by the worms, but both are due to a common cause—a diet wanting in sulphur, lime and salt. It is only in the very first days of the child's life that the milk should be diluted with water or gruel, if child mortality is to be diminished. In passing we would state, that the tape-worm (in opposition to the received view that it originates from the flesh-worms of the pig,) is really produced from the phosphatic epithelium of the intestines. Thus in cases of persons whose blood is poor in earthy constituents, as also in the case of women in pregnancy, who are obliged to provide the materials for a second organism, we often notice a desire to eat chalk, ashes, or the bark of trees. The bark of trees leaves a large quantity of ashes rich in lime. The human organism experiences a desire for what it requires. As the giant Antaeus constantly obtained fresh strength, whenever he came in contact with his mother earth, so we also gain fresh strength from earthy substances, while their absence gives rise to morbid conditions.

A Parisian Professor who was in the habit of regularly spending his holidays by the lake of Geneva and while there consuming large quantities of fish, always came home with a tape-worm! From this he concluded that the germ of the tape-worm must be parasitic in the gills of the fish, whereas it was due exclusively to the limeless phosphatic flesh of the fish which was liable to metamorphosis into tapeworms, which often occurs in the case of fishermen. In addition there are not wanting proofs to clearly show the difference between the conditions of existence and sustenance of ordinary worms and of the earth-worm which calls himself Man!

By way of example, I will call attention to our beet-root plantations. The country from Magdeburg to Brunswick for a long time gave such crops of sugar-beet as to cause numerous sugar manufactories to be built there. The fertility of the soil was due to the *débris* of the porphyritic rocks of the Harz*) of which it is formed, which contains sufficient quantities of silicates of potash, soda, lime, and magnesia. As these constituents of the soil became exhausted by the long continued cultivation of beet-root, the plants were unable to obtain the materials required for their normal growth and their substance degenerated into worms (Nematodes). This evil was not abated until manuring with nitrates (chili-salpetre) was resorted to, although it would have been more sensible to cart powdered porphyritic rocks on the fields.**)

From the facts now mentioned we may manifestly deduce the law, that the earthy constituents, potash, iron, magnesia &c. together with

---

*) *Geologically* it would be perhaps more correct to say that the soil of the North German Plains is derived from the "boulder-clay" of the glacial period, but as this is also the product of the attrition of the igneous and crystalline rocks of Scandinavia the *chemical* aspect of the question remains the same. Tr.

**) Here also cause and effect were mixed up, the worms being regarded as the cause of the beet not flourishing. Thus M. Girard who in 1890 was charged by the French government to investigate the cause of the rise of beet-worms, stated that the worms hid in the soil during winter, attacking the plants in the spring. He also held that the beet worm was eaten by animals, that the worms passed through their alimentary canal uninjured by the universal decomposition of the contents of the intestines, multiplied in the dung, and were again carted with it on the fields. No one as far as I know has protested against this absurdity.

their salts increase the electric tension in certain forms of life and so lengthen life, provided cool air, rich in oxygen be not lacking.

Sweden and Norway afford interesting examples of the extent to which human life may be lengthened by a diet rich in ashy or earthy constituents, accompanied by the respiration of cool, pure air.

In the *"Christiania Dagblad"* for 1884 we read various accounts bearing on this question:

"Beret Olsdatter of Ravneberg will celebrate her 103^d birthday on Dec. 1. having been born Dec. 1. 1781. She has had 10 children of which 8 are still alive; the oldest is 83. Her husband died 14 years ago. They lived and kept house at Faaberg. Such were the conditions in which they lived that they regarded half a ton of corn as a rich harvest. The husband was compelled in 1814 to march against the Swedes. That year the whole harvest was frozen, so that corn could not even be bought. She then lived by scraping bark from trees, and gathered moss and ground the whole to flour. Industrious as she was, she always succeeded in this way in getting enough to eat, so that many said to her "I don't understand how you live". She was a good needle woman and could weave and spin, which assisted her in getting a living. Later on she and her husband took to fishing. The piety of the family overcame all difficulties. Her memory she preserved in all its freshness up to her present very advanced age."

This account is interesting as showing that the natural judgment directed the hungry woman to seek for nourishment from tree-bark, which in comparison with wood contains on an average 10 times the amount of earthy constituents especially lime.

In addition the same paper of Dec. 16 1884 No 450 contains the following:

"On Dec. 7. the oldest inhabitant of Stockholm Fru Katarina Julin celebrated her hundredth birthday. She has been a widow 34 years. During the last 25 years she has not left the small room in which she lives. She is still in possession of all her faculties."

In Sweden the custom prevails of mixing silica-meal with flour for baking, and it appears, as if this had the effect of imparting greater firmness to the muscles; at any rate the meat of healthy cattle leaves an ash on burning containing 2% of its weight of silica.

Finally the "Dagblad" for Sunday Jan. 4. 1885 says:

"On New Year's Day in Laerdal was celebrated the diamond wedding of the Gunsmith Lasse Senjereim and his wife Ragnhild Olsdotter. The husband is about 92 and the wife 85. They have had 8 children 7 of whom are alive, 4 being in America and the others here. The old man is lively and active, and still works at his smithy in spite of his 92 years."

It may be seen from these reports all occurring within a few weeks of one another that the cooler Scandinavian climate affords a better chance of a lengthy life than do the warmer regions of our planet. It even appears as if the concentrated electric tension of the Scandinavian granite passed over into the men who move upon it. Granite consists of silicates of potash, soda, lime, magnesia, iron, and alumina. It has accordingly the chemical properties of a glass flux, which has the physical properties of holding the electric fluid together, whereas a warm moist coast climate carries off the proper bodily electricity, causing a greater amount of decrepitude and a shorter life.

As far as the granite is concerned that of the "Giant Mountains" of Germany proves itself the equal of the Scandinavian variety. And in the "Giant Mountains" we also find persons of great age with a meagre average diet. There is therefore some reality in the transmission of electricity from the rocks to the human body.

Having thus in general considered the effects of cool air for respiration and the electric tentional action of salts and earthy bases, we will proceed to acquire a still more accurate knowledge of the chemical processes, which cause the dissolution of the human body when there is a lack in the three conditions before mentioned, by the consideration of a number of special cases supposed to be of peculiar difficulty I select for this climatic fevers, smallpox, consumption and epilepsy. When we have grasped the chemical laws by dealing with these important diseases, we shall not have much difficulty in discovering the remedies for the other ailments, which are only lighter varieties of these four diseases, respectively affecting the vascular system, the lymphatic system, the connective tissue and the nervous system.

B. SPECIAL PART.

A well known proverb says: "Nothing can come out of the sack but what has been put into it." Now since physiological chemistry teaches us that the final product of the combustion of our bodily substance in the oxygen we breathe is dehydrated ammonium carbonate ($N_2 H_4 CO$), to which the name of urea has been given, as it was first discovered in urine (which removes the burnt up constituents dissolved in water from the circulation), it is plain that this same urea, combined with combustible substances must form a basis of our tissues, and this conclusion is supported by facts.

The combustible materials which in combination with urea go to form both our muscles and the gelatine of our connective tissue are cyanogen (CN) and its hydrogen compound (HCN) and acelytone (CHH). In certain compounds the dehydrated carbonate of ammonia (urea) exists as such or in the isomeric form of ammonium cyanate ($COHN$, $NHHH$), for example in the ammonium salt of carbamic acid. (cf. p. 63).

In other compounds e. g. xanthin, sarkin, and creatin it is recognizable in the unoxidized condition namly as gelatine-sugar or glycocoll ($CHH COO NHHH$). Owing to the occurrence of tissue in regular layers this glycocoll or "gelatine-sugar" must be regarded as existing in masses; thus the fact that urea is produced from 2 equivalents of glycocoll is explicable from the regular oxidation of the latter with 6 of oxygen. In this case one double molecule of glycocoll gives rise to 1 of urea, 3 of carbonic acid, and 3 of water.

Sarkin the basis of flesh would give rise in complete oxidation due to normal respiration to 1 of urea, 4 of carbonic acid, 2 of water, and 3 of nitrogen. These 4 products of oxidation are to be regularly met with in venous blood, so that a good deal of the nitrogen in the expired air is not due to the nitrogen of the air inspired, but to the oxidation of tissue in the body itself.

If now instead of complete oxidation merely an incomplete and insufficient combination with exygen takes place, it is plain that some half burnt products, so to speak, of the combustible substance are produced, similarly to the way in which when the door of a stove heated with wood is closed during the burning, the next morning instead of white ashes we find black charcoal. Similar half-burnt compounds produced in our bodies by incomplete respiration will be the cyanogen compounds which are oxidized with more difficulty than the hydro-carbons. But unburnt cyanogen cannot be detected in the blood as such, since it instantly combines with the iron of the hemoglobin of the blood to form prussian blue. In this manner small quantities of cyanogen are rendered harmless. It is only when the iron of the blood bas been in this way largely used up, that the absorption of oxygen decreases, and as the "cyanosis" increases, the difficulty in breathing, the oppression of the heart and a general feeling of weakness become noticeable, as it necessarily must, since the feeling of strength is due to the respiration of oxygen.

The amount of oxygen required to produce a certain amount of strength or to perform a certain amount of work may be estimated with comparative accuracy. Whether we consider the hydrocarbons contained in creatin, gelatine, or fat as the combustible material, it follows that about 160 grammes of hydrocarbons are consumed in a day's work in a temperate climate. For this quantity nearly 3 1/2 times as much oxygen in weight would be required, i. e. 560 grms. to oxidize it to carbonic acid and water; such a quantity of oxygen would be equivalent to 5000 litres of air at the winter temperature of zero Centigrade from which follow two things:

Firstly, that we must of course replace the 160 grammes of hydrocarbons which we have used up in the form of nerve-fat, blood-gelatine and muscle-substance, and also the ashy constituents with which they were combined (phosphate of ammonia, phosphate of lime, phosphate of potash &c.) if we desire to be able to expend a similar amount of energy on the following day; and in the second place, if we work in hot air, which is perhaps ten times as thin as cold winter air we would be obliged to respire 50000 litres of air daily to obtain the necessary supply of oxygen; our lung-bellows are not, however, formed to perform this amount of work, so that we must fall back upon some other means of assistance.

Unfortunately the replacement by nutriment during the respiration of hot air does not keep pace with the loss, since to keep the digestive processes going properly an adequate supply of oxygen for the abdominal nerves is requisite; therefore the hot (rarified) air in the tropics causes that sudden atony known as climatic fever.

We will now proceed to trace what fate befalls the hydrocarbons of our body which are combined with nitrogen.

CLIMATIC FEVER.

I made the statement at the beginning of this treatise that chemistry can solve the apparently most difficult problems by simple numbers and equations. This I will proceed to prove by selecting a subject from the domain of pathology supposed to be particularly dark and difficult, namely, *yellow fever*; for the successful treatment of which disease the Government of the United States of North America is said to have offered a prize of considerable value.

Yellow fever is met with along the coasts of tropical countries and usually ends in death in from 3—7 days after the first chill.

It commences with headache, sleeplessness, weariness, want of appetite, oppression of the stomach, and fluctuating pains in the back, loins and limbs. From this it may already be observed that the circumstances forming as it were the introduction to the specific disease, are not independent phenomena of a disease, but merely connected symptoms, springing from one and the same source and such as are also characteristic of anaemia.

After the lapse of a certain time which may be a few days, or several weeks, shivering begins. Thereupon follow increased headache, restlessness and insomnia. Lying in bed becomes unbearable. Rising in the stomach and bilious vomiting follow, and the skin becomes yellow.

This yellowness of the skin grows more and more intense; passing from lemon color to saffron; dark brown spots being observeable here and there. The tongue then becomes dry and crusted over; in many cases it turns perfectly black and trembles when the patient speaks. The vomit consists of corrupted blood and a black substance like the sediment of coffee. The stools which take place involuntarily, consist of a black tar-like substance derived from the blood.

The lips are dry and crack, and the amount of urine secreted is small, sometimes there is none at all.

Weak pulse, cold skin, and thick, clammy perspiration complete the specific symptoms of the disease. In addition, ravings like those of typhus, and delusions of the sight and hearing like those of delirium tremens occur, and finally there are cramps followed by death.

If there is recovery, it takes place through a quiet sleep and the breaking out of a warm and copious perspiration on the seventh, ninth or eleventh days.

These symptoms, apparently so complicated resolve themselves into the following simple physical and chemical processes.

According to Prof. Grimaldi's experiments, water vapor absorbs 7937 times as much heat as dry atmospheric air. Now since the air on the hot coasts is laden with moisture, the heat and the moisture together act on the one hand in such a way as to *carry off electricity* (and we have fully shown that it is electricity which holds our bodily substance together), while on the other hand it decreases the proportion of oxygen in the air.

These two facts contain the first causes of yellow fever, and from them all that follows may be explained.

Parallel with the decreased amounts of oxygen respired, an accumulation of carbonic acid in the blood takes place. This is due to the fact that expiration and inspiration stand in a certain fixed relation to one another. If the inspiration be feeble the expiration will be feeble also, since the lobes of the lungs always retain a certain amount of air. Now the longer the inspired air remains in the lungs without being expelled by fresh air, the more fully and completely it loses its oxygen, and the more the carbonic acid which is produced by the combination of the oxygen of the blood with fatty and gelatinous substances and the substance of the muscles, accumulates in the blood.

Now since the carbonic acid which has formed in the blood, expands in agreement with the rarefaction outside, like as the gas expands inside a balloon as it rises, it follows that the roomy cranial veins (the Sinus longitudinalis superior, Sinus rectus, Sinus cavernosus, Sinus petrosus superior and inferior, Sinus transversus and occipitalis) suffer most appreciably at first. Hence the *pains in the head.* For the brain-substance is everywhere pressed upon by the expanding carbonic acid and consequently the circulation in the fine capillaries is impaired.

The same fate befalls the venous plexuses of the spiral column (plexus spiralis externus and internus) likewise the systems of the vena azygos and hemiazygos, which follow the course of the spinal column. Hence the *pains in the back.*

The large veins of the thighs (*vena cruralis*) also participate herein. Hence the pains in the loins. The system of the portal veins is of course also affected, and where the abdominal regions affected are connected with branches of the cerebral nervous system (pneumogastric), as is the case in the stomach, this manifests itself in *oppression of the stomach.* By the same cause namely stagnation of the blood in the capillaries, in so far as owing to insufficient introductions of oxygen into the blood the circulations is disturbed, are to be explained the *pains in the limbs* occurring first in one place and then in another.

In addition, the inadequate amount of oxygen in the blood causes the *nerve-fibrils*, which only perform their functions in proportion as a certain amount of their substance is consumed, to become gradually paralyzed. Neither the cardiac plexus nor the intestinal nerves receive a sufficiency of oxygenated blood. Hence on the one hand *weaker pulse* and on the other diminished secretion of the digestive juices (gastric, intestinal and pancreatic secretions). The result is loss of appetite. And since nothing is eaten, the absorption of chyle also ceases. The liver, indeed, still continues to produce bile from the blood of the portal vein, but the bile is not as ordinarily put to use in the production of chyle because there is no introduction of food; and consequently the bile enters the duodenum, passes thence into the stomach and is thrown up.

The discontinuance in the formation of fresh lymph gives rise to increasing weakness, dejection and exhaustion.

The kidney nerves (the renal plexus) also cease to act from want of stimulating oxygenated blood, and the *secretion of urine ceases*. In consequence of this the urea remaining in the blood takes up water and is changed into ammonium carbonate, the direct effect of which is to paralyze the nerves. Hence *prostration*.

The *turning yellow of the skin* is explained as follows.

The ammonia and the ammonium carbonate arising from the decomposition of urea attract water, and this water is drawn from the albumen of the blood. The cyanogen liberated from the muscles acts in the same way.*)

We will now proceed to enquire what chemical metamorphosis gelatine undergoes when it is imperfectly oxidized and in addition is chemically dehydrated. That very poisonous products may be formed from gelatine by the presence of putrid substances under exclusion of air, is well shown by Koch's lymph (tuberculin), which is made with the exclusion of air. We will at once proved to explain how this is possible.

Gelatine may be decomposed by sulphuric acid into leucin and glycin. These two constituents would on being completely oxidized by 21 parts of oxygen produce carbonic acid, water and nitrogen. When, however, insteat of this only 3 parts of oxygen come into action, producing 5 parts of water and 1 of carbonic acid, the result is the formation of prussic acid and a yellow aniline oil, which latter may also be produced by the dry distillation of bones, which also contain gelatine.

The computation is as follows:—

---

*) A Berlin carrier of letters was poisoned with potassium cyanate given him in a glass of spirits. He staggered down stairs crying "For God's sake water" His murderer, who when he thought he was discovered, had taken the same poison, uttered the same exclamation. So parching is the effect of the water-attracting cyanogen!

Leucin	$= C_6 H_{13} N O_2$
Glycin	$= C_2 H_5 N O_2$
	Total $C_8 H_{18} N_2 O_4$
$+$ 3 oxygen	$= O_3$
	$C_8 H_{18} N_2 O_7$
Deduct 5 particles of water and 1 of Carbonic Acid	$= C H_{10} O_7$
	Remainder $= C_7 H_8 N_2$

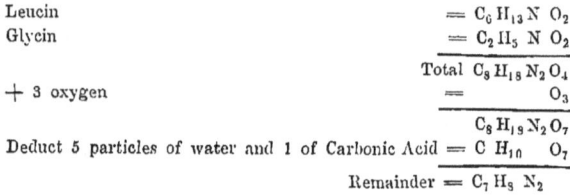

.This Remainder contains the elements of prussic acid HCN, and aniline oil $C_6 H_7 N$.

Aniline oil colors the skin intensely yellow, while prussic acid turns the iron of the blood into blue-black prussian-blue.

This dark blue-black product which is incapable of chemical combination with oxygen, appears visible through the fine capillaries of the tongue, the epithelium of which is uncommonly full of small vessels. Furthermore the trembling of the tongue in speaking is due to want of oxygenated blood.

The formation of the dark blue cyanide of iron and potassium (Prussian-blue) also occurs in the capillaries of the mucous membrane of the stomach and intestines; from the pores of these capillaries the carbonic acid which has accumulated to excess in the blood, issues forth in gaseous form carrying some black blood with it. Hence the vomiting from the stomach of black masses and the unconscious mechanical passing of black tarry masses of blood from the intestinal canal. The tarlike form of the blood is caused by prussic acid and is also characteristic of other cases of poisoning by prussic acid.

As the amount of active oxygen continues to grow less, the nerves of the skin grow cold, and hearing, seeing, and thinking become confused; for also the brain needs oxygenated blood in order that it may perform its functions normally. Finally the patient lapses into apathy, and death ensues.

Those persons are spared by the epidemic who with the coming of the hot season visit the neighboring mountains, as the heavy moisture-laden air of the plains does not reach up to their height, and owing to nocturnal radiation they enjoy a nocturnal *change of temperature* from which the *"electric fluid"* is regularly formed. Correspondingly the epidemic dies out as soon as a thunderstorm breaks in upon the equable heat which has lasted for months, since this provides the atmosphere with fresh electricity which passes over into the human organism.

Such a state of affairs shows us plainly enough the course that therapeutics ought to pursue to successfully combat yellow fever. We will deal with that in the chapter treating of therapeutics, but will first proceed to the consideration of other climatic affections.

Among the peculiar symptoms of the diseases due to the influence of weather and climate are especially: Dyspepsia, dysentery, vomiting and

diarrhoea, jaundice and intermittent fever. This whole circle of phenomena is unmistakably characterized as caused by paralysis of the intestinal
nerves which regulate the functions of the various glands. This paralysis is
physiologically due to want of oxygen in the blood and to its being
replaced by carbonic acid and ammonia, in the atmosphere of which
the nerve-fat is not oxidizable, just as a lighted candle will not continue
to burn in it.

In marshy districts, such as Holland, where dense vapor presses on
the ground; in fortresses surrounded by moats containing stagnant
water, and in plains in the neighborhood of bodies of water with mountain chains behind them impeding the movement and dissipation of the
exhalations, as is the case at Hoboken on the Hudson with the Jersey
Heights behind; further in tropical regions during the wet season or
when swarms of mosquitoes and flies hatched from the marshes fill the
air, the atmosphere is as poor in oxygen as it is in electricity. For as
already pointed out, the layer of vapor naturally presses upwards the
dry and consequently electric layers of the atmosphere and raises them
above itself. It is due to this fact that both mountaineers and those
dwelling in the upper stories of our modern stone palaces are more
secure from diseases than those living on the ground floor who are
exposed to the operation of the moist ground which conducts away
electricity.

The loss of electricity originally affects only the nerves of the skin,
since electricity naturally collects on the surface of the body whence
it radiates; but since the nerve material forms an integral system, the
condition of the external nerves is soon communicated to the nervous
material of the inner surfaces and consequently to all the glandular
organs. The principal of these is the inseparable quartet: liver, stomach,
pancreas and spleen, on which digestion and nutrition depend.

Owing to the stoppage of the circulation in the spleen where the
arterial capillaries are separated from the venous capillaries, the electroscopic phenomenon of shivering takes place, and it is followed by
various processes of decomposition which are partially to be ascribed to
the ammoniacal disintegration of the blood clot which accumulates in
the spleen, which liberates a certain amount of heat which we feel as
fever-heat.

But even without catching cold from without, without a rainy
season and without the presence of marshes, merely through the
continuance of uniform hot weather even in dry high table lands as in
Peru, fever may strike root. This is due to the fact that our organism
during the hot period of the year attempts to equalize the external with
the internal temperature by evaporation which absorbs heat, perspiration
breaking out from all the pores. The elimination of perspiration is,
however, antagonistic to the secretion of urine; the one excludes the other.
And when the secretion of urine ceases or is impeded, a certain

proportion of urea remains in the blood and decomposes into carbonate of ammonia with the consequences already mentioned namely: the absorption of water, the formation of prussic acid, the killing of the hemoglobin which absorbs oxygen (cyanosis), the swelling of the connective tissues through the water which is secreted (hydrops), and the aggregation of the dying blood corpuscles in the spleen causing it to swell. Bloatedness, paleness of the face, and apathy accompany one another and steadily gain ground. Such are the characteristics of marsh fever (malaria) which in contradistinction to yellow fever might be called white fever.

Yellow fever is endemic in the equatorial regions, replete with watery vapors, on the west coast of Madagascar where the mountain range that traverses the island and the mountainous ridges on the opposite coast of Mozambique, in consequence of the rotation of the earth from West to East, impede the escape of the watery vapors that have been formed under the equatorial sun, so also on the Gulf on Guinea where the steep mountains prevent the escape northward of the masses of vapors raised from the sea. In addition it rages along the flat coasts of the gulf of Mexico up into the broad valley of the Missisippi, where owing to the revolution of the earth the vapors are banked up against the western mountain ranges and also along the East coast of North America as far up as Philadelphia; in these cases the Rocky Mountains, the Apalachian and the Alleghany Mountains obstruct the escape of the heavy watery vapors. While we have thus delineated the chief regions where yellow fever rages, there are similar conditions in Asia— along the coasts of the Bay of Bengal where the high mountains of Thibet present an insurmountable wall nearly half a geographical mile in height, to the vapors which rise from the Indian ocean—here we encounter the Indian Cholera.

The causes of these climatic affections are always to be found in hot air poor in oxygen on the one side, and on the other in heavy watery vapor in the absence of thunderstorms that would precipitate the moisture. In our latitudes the moist *autumn air* which carries off electricity, lying heavy on the ground may also produce cholera.

This native cholera (*cholera nostras*) only differs from the Indian, in being a milder variety just as the bilious fever of our climates differs from the tropical yellow fever only with respect to the intensity and rapid development of the latter.

Our dependence upon atmospheric conditions shows itself under the most various conditions. For example men suffer for 24 hours from headache and sleeplessness before the „Föhn" (the hot south wind in Switzerland) arrives. In damp and musty dwellings we fall a prey to crick in the neck and rheumatism, and foggy air produces epidemic catarrh (Russian Influenza?). On the other hand yellow fever cannot reach 3000 ft above the sea-level, since there the peculiar electricity of air in motion is communicated to our bodies. The truth of this proposition is shown even in the case of lifeless flesh. Thus while on the coast of Brazil meat must be boiled or roasted on the day the animal is killed, to prevent its spoiling, and for the same reason human bodies must be buried on the day of death, on the other hand meat hung out to dry on the high points of the Grison Alps or the Cordilleras never decomposes at all.

We are all healthier in a dry atmosphere on a sandy soil which holds the electricity, than we are in moist regions which carry off the proper electricity from our bodies.

When we study the case of the city of Cairo, situated in the flat coast district of the Mediterranean, and deprived of electricity as it is under certain conditions of weather and in certain years by the basin of the Red Sea which being enclosed by mountainous ridges collects a mass of vapor, thus giving rise to cholera; and when further we see in the Bay of Bengal, where the conditions are still more unfavorable both plague and cholera breaking out, there is really no need to hunt with the microscope for a "bacillus" as the cause of cholera, and to steam from Alexandria to India for the purpose of trying whether the Alexandrian bacillus can be transmitted to Indian Apes.

There can be no doubt even *à priori* about the possibility of such communication, as a whole tub full of dough can be set in fermentation by a handful of leaven, when the temperature is sufficiently high. But of this sort of communication there is no question as it is not customary to inject into a healthy person the blood of one suffering with cholera.

As is the case with yellow fever, cholera disappears as soon as there is a plenteous fall of rain. The explanation of this is, that the water in passing from the gaseous to the liquid condition sets free considerable quantities of electricity, which become available for the atmosphere and the men living in it. The apparent converse of this produces the same effect, since the evaporation of large quantities of water in hot air renders a great amount of heat latent. The consequent cooling of the atmosphere is accompanied by the appearance of electricity since the latter is due to variations of temperature.

The following notice from the papers for Aug. 8. 1883 is quite in accordance with this conclusion:

"This year's overflow of the Nile is quite exceptional. Since it has occurred the number of cases of cholera has undergone a rapid decrease.

It is obvious that in the same way as in the glands different substances are secreted from the same material i. e. the blood, so modifications of the factors which are active in earth, air and water, must give rise to various products of decomposition capable of impressing a particular character on bodily ailments. The intensity of the atmospheric influences plays an important rôle in this respect. For instance it cannot be without importance whether a certain degree of atmospheric heat be divided over 7 or over 14 days. As a striking example in point, I would adduce the production of electric light. We are aware that when the velocity of the dynamo falls below a certain point no electric light at all is produced. It is only when the velocity reaches a certain degree that interaction between electricity and magnetism or the transition from the one into the other is reduced to moments of such close proximity that light may me born therefrom. This analogy enables us to understand how the swifter rotation of the earth in the equatorial regions causes a different disposition of our organism from that prevailing in a northern latitude of say 35⁰.

We have thus seen, that the specific disease known as cholera is characterized by a decomposition of the blood into watery lymph and fibrous tissue, and this as discussed on page 24 is an electrolytic process.

Owing to the coagulation of the fibrin in the fluids of the tissues the chemically detached lymph is *mechanically forced out* through the walls of the stomach and intestines giving rise to diarrhoea (discharges like rice-water) and vomiting. Of this contractile power of the coagulation of fibrin there are several proofs. Thus for instance we are aware that in all corpses, owing to the weight of the lower jaw the mouth at first opens, but it gradually closes, a mechanical effect of the contraction of the facial muscles caused by the *rigor*. This universal observation is supplemented by a very peculiar phenomenon which has been noticed in the corpses of those who have died of cholera. It has happened a number of times that a very busy doctor in time of cholera epidemies would sign the death certificate as soon as the heart had ceased beating and then would leave. As soon as he had come home, however, some one of the relations of the deceased would rush in all out of breath and almost beside himself and cry out: "O doctor, our father bas come to life and lifts up his arm and moves his fingers".

To understand this strange phenomenon we must remember that in any dead body without exception when the muscles of the arm are exposed and the proper tendons are pulled, the fingers may be made to move either separately or all together like the keys of a piano. This shows that the bending or straightening of the fingers as also of the arm may take place without the action of the will which during life effects the muscular contraction.

A further phenomenon which occurs in the case of healthy patients will make this matter still more comprehensible. Workmen on streets such as stone-setters and pipe-layers when indulging in their necessary mid-day sleep frequently place the left fore-arm under the body and lie upon it. In this case one of the two subclavian arteries which carry the blood from the aorta to the right and to the left to the upper arms, has the circulation from the elbow down-wards checked by the pressure of the body upon the for-arm. After a certain time, since the pulsation of the heart has been accustomed to drive the blood into both arms equally, it doubles its effect in the free arm, the result of the double supply of oxygenated blood being of an almost explosive kind. The free right arm is suddenly jerked upwards without the exercise of the volition of the sleeper, or his being able by his will to prevent it. He then usually awakes, changes the uncomfortable position of his arm and continues his sleep. Any one passing by, however, would suppose that the sleeper had been dreaming.

Now since in cholera corpses, even after the stopping of the heart and the respiratory organs as also of the brain functions, jerks of the tendons still occur as above described, it is plain that a continuing separation of lymph (serum) and fibrin is going on, since only fibrin when produced in large quantity, can by its coagulation, effect a change of volume and thus a shortening in a mechanical manner.

There is nothing which sheds so much light on the real processes in cholera, as this hitherto not sufficiently noticed post mortem movement of the limbs. All the other symptoms are only rendered comprehensible by the accumulation of fibrin and the loss of watery lymph.

Among the other symptoms of cholera the cramps in the muscles of the calves of the legs are specially noticeable. The phenomena of strongly diminished secretion of urine, mental indifference, weak pulse, diarrhoea and vomiting and striking coldness of the skin have less specific importance, as they also occur in other diseases.

As far as the cramps in the calves are concerned by which the feet are involuntarily drawn up to the body, they are to be explained like the muscular movements which occur even after death, by mechanical tractional force exercised on the muscles by the coagulating fibrin. In correspondence with this, it is found that the clammy perspiration which is pressed out from the pores contains albumen i. e. lymph. The small amounts of urine also which are secreted, contain albumen, as the blood owing to loss of lymph becomes so thick that it can no longer give up water, but only albumen. Blood thickened in this way can only with difficulty be set in motion by the heart. Hence the thin, threadlike scarcely perceptible pulse. And as no more lymph reaches the brain, the mental energy rapidly decreases. The diarrhoea and vomiting are due to the same cause as the clammy perspiration; the lymph is forced out of all the serous membranes set with glands, in the same way as it is from the sweat-glands on the surface of the body.

If any doubts be left as regards the effect of diminishing the vital forces, caused by this intensive chemical separation of the albumen of the blood into watery lymph and fibrin, the law of the unity of all natural forces, will enable us to remove them. For from this law it follows that the same amount of heat disappears on the decomposition of a product as was necessary to form it. Now there is no disease which causes the patient to become so deadly cold as cholera, just because the re-arrangement of the molecules of the albumen of the blood uses up an enormous amount of heat.

The deadly processes of cholera would be to a certain point sufficiently explained by the consideration that the vital force must sink down in proportion as the amount of albumen in the blood decreases, somewhat as it is with animals slaughtered by withdrawing their blood. However, this explanation is not complete, because in ordinary cases of loss of blood e. g. from wounds in the battle field, the propelling power of the heart is diminished by a sort of automatic regulation. The blood flowing slower and slower from the wound, coagulates, the wound closes, and the blood being renewed and increased from the lymph, convalescence takes place. In cholera, however, nothing of the kind takes place. We have on the contrary also to deal with a genuine case of *blood poisoning*. This is indicated by the blue colour of the lips and hands, due to prussiate of iron which admits of no other conclusion than that it is a poisoning due to prussic acid spontaneously generated by a chemical disintegration of albumen, simultaneously with the decomposition of the albumen of the blood into lymph and fibrin.

Certain occurrences in times of cholera, such as that persons hitherto healthy suddenly fall down dead in the street, can only be explained as the result of internal generation of prussic acid, due to the effect of a peculiar atmospheric electricity.

If it should be enquired, how it is that in cases of cholera such comparatively small amounts of prussic acid should destroy the system, I would draw attention to two facts of a similar kind. The one relates to a much used product of albumen which is formed by allowing barley grains to sprout, when they divide their albumen into two unequal parts—one of the products being *fibrin* rich in glycocoll, to form the root, and the other a sweet *lymph* containing but a small amount of glycocoll but very rich in grape-sugar. This sweet substance is termed *diastase*. Diastase possesses the property of converting, at a temperature slightly above that of our blood, quite astounding quantities of starch if this has been before cooked into a paste into the same sugary substance contained in the barley-malt, namely into *Maltose*. This we may suppose due to the fact that the mere approach of the diastase through its electro-chemical opposition to amylum (starch) effects the decomposition of the latter into sugar by the absorption of water. When this decomposition has taken place, fresh portions of dextrin come into contact with the diastase and share the same fate. The only limit to this action is the destruction of the characteristics of the diastase by the oxygen of the air.

We may assume the same in regard to prussic acid which on combining with albumen acquires the property of converting considerable quantities of blood-albumen into similar prussic-acid from albumen, unless a limit is set to this action by the oxygen of the arterial blood.

This latter is generally the reason why small amounts of prussic acid are not fatal.

When we consider that starch is produced from 6 molecules of sugar by the detachment of 6 molecules of water and that it can become sugar again by reabsorption of 6 molecules of water, the decomposing power of the barley-albumen becomes as easy to understand as the way in which 6 cords of wood can be split up by one and the same word chopper. The wood chopper gains fresh warmth and strength with every breath he draws, and probably the barley-albumen absorbs as much water on one side as it gives up to the starch on the other. Accordingly the opposite side splits up new material, while that which has been deprived of water redintegrates itself from the starch-paste owing to the warmth. In this way the process by which a substance can continue to decompose without other means than the constantly renewed application of heat, becomes comprehensible. From this we see again, that Mayer's law is without an exception.

All these facts, I think, prove that atmospheric influences first cause the appearance of cholera. I will now proceed to throw light on certain superstitious opinions.

It has been found that cholera and yellow fever only spread as far as a man can walk in one day, and that they extend with preference in those directions in which a number of persons are in the habit of passing, i. e. along the high roads. From this it has been concluded, that cholera is communicated from one person to another, and letters from cholera districts have been disinfected and quarantines established for vessels coming from such countries.

With regard to this we may observe: That particular states of the atmosphere depending upon the sun can only gradually extend over neighboring countries, since in so far as sunshine prevails at one place, cool shade must be found at another as a matter of course. It is also certain that cholera prefers to travel along high-roads. But this is due to the fact that the high-roads are made as level as possible to facilitate the drawing of great loads with the least effort, i. e. they are constructed in the valleys. And what do we find in the valleys? As a rule rivers. And rivers, the vapors of which during hot summers collect in the valleys conduct off electricity. This is the reason why yellow fever extends all along the flat valley of the Mississippi from New Orleans to St. Louis. And the valley of the Ganges from Delhi to Calcutta forms an equally favorable district for cholera.

It is *no* superstition that dread of infection disposes to cholera; for all emotions induce a using up of electricity and so diminish our powers of resistance. It is therefore quite right that cholera patients who have no sick room of their own, suitable for their reception should be removed to a hospital to avoid causing sickness through loathing and fear with those who surround them. The necessary directions for the treatment and the prevention of cholera will follow in the therapeutical part of this work.

SMALL-POX.

While we find in the cases of cholera and climatic fever that death results from a more or less extensive disintegration of stagnating blood, small-pox has its origin in the contents of the lymphatic system, i. e. in the lymph.

The lymph is intended as already explained to supply the blood and the nerves with fresh material to replace what has been used up. It is in the first instance sucked up from the intestines in the form of chyle i. e. a milky mixture of bile, fat, a food-extract, collected by a number of fine canals in the "receptaculum chyli" or the lower portion of the thoracic duct which passes up along the vertebral column, emptying into the venous system near where the left internal jugular meets the sub-clavian vein. The base of the lymphatic canal is widened near the second and third lumbar vertebrae into the "receptaculum chyle" from which pass two branches, each one of which dissolves into a netform plexus from which smaller vessels pass to supply both sides of the *pelvis* and the *thigh* with fresh material.

A similar duty is performed by the upper part of the main lymph-canal; this sends branches on one side to the left and lower portion of the thorax as also to the left half of the neck and head together with the left arm, and .on the other side to the right upper half of the thorax, the right half of the neck, the right half of the head and the right arm. The lymph that is not used up in this course, passes into the stream of blood between the right internal jugular and the sub-clavian veins. This anatomical relation of the parts throws considerable light on the regions where the small-pox pustules make their appearance.

The course of the thoracic duct between the Aorta and the Vena Azygos along the vertebral column and consequently along the spinal marrow, enables us to see that both the circulation of the blood and the electric currents passing through the nerve trunks produce an effect on the progression of the lymph; for whatever acts side by side, also acts reciprocally.

A stagnation in the movement of the lymph may accordingly take place when the blood is deficient in iron and salts which act electrically on the nerve trunks. In addition we must take into consideration that the amount of alkaline and earthy salt in the blood normally only reaches 1%, whereas the amount contained in the lymph extracted from the chyme under ordinary circumstances reaches half as much again viz ($1\,1/2\;\%$). Such being the state of affairs it is plain that when the amount of salts in the lymph is unduly diminished and sinks below the percentage contained in the blood, it becomes physically very difficult for the lymph to empty itself into the blood, as a lighter liquid will not sink in a heavier one and the lymph which is held back consequently stagnates in the vessels, a state of affairs which must the more surely give rise to chemical disintegration as this same weak lymph is not sufficiently protected by salts against putrefaction.

In agreement with this explanation of the origin of small-pox we find that it particularly attacks children that are given diluted cow's milk instead of the natural mother's milk,—the amount of alkaline and earthy salts contained in the former often falling short of 3 parts in 1000 as against 10 parts in 1000 in the mother's milk; since cow's milk has on an average even when undiluted only 6—9 parts per 1000, This deficiency in salts ought under all circumstances to be made good when cow's milk is used for infants. That the lymphatic system is in reality the seat of the disease is fully shown by the whole course it pursues.

The stagnation makes itself first perceptible near where the lymph should enter the circulation, namely in the region of the throat and the adjoining regions of the face and breast, where numerous lymphatic glands occur. Here it is that the pustules make their first appearance emptying the lymph outside, because the blood has no use for such strengthless lymph.

As the disease progresses the pustules appear also on the abdomen. the arms and the legs, and first of all in the neighbourhood of those places which are characterized by an abundance of lymphatic glands, in which in any case the movement of the fluids is liable to be impeded.

The characteristic pains in the pelvis and loins as also in the back, point out the thoracic duct as the seat of the evil, passing up as it does along the vertebral column and sending branches to the pelvis and thighs.

The fact also that the disease as it progresses passes over into the serous membrance of the brain, and the serous pleura which convey lymph, is explained by the anastomosis of the lymphatics which penetrate all the tissues of the system forming a complete and united network.

When further more blindness is often found to be among the effects produced by small-pox, this also only shows the continued extension of the disease to all tissues conveying lymph, as it finally extends even to the vitreous humor of the eye, which with the exception of some minute membrances slightly gelatinous, consists of pure lymph.

Everything points to the lymphatic system as the origin of the affection, and the comparatively slow extension of the disease accords with this view.

Cholera, and cholera morbus, originating in the blood, which in a few minutes courses through the whole system, develope with 'great rapidity; on the other hand the system of lymphatic tubes as the region in which the small-pox originates, causes a very much less rapid progress, corresponding to the slow flow of the lymph, which is impeded by innumerable valves and requires even in a normal state 12 hours before it is completed.

Although as has been said, the developement of the disease takes place in the lymphatic system, blood and nerves, nevertheless, participate in it very considerably.

In so far as pains in the lumbar region and along the back precede the characteristic signs of disturbed digestion (pains in the stomach, nausea, vomiting, coated tongue &c) it would appear that the nervous system and, indeed, the pneumo-gastric nerve is the first affected. As, however, the proper action of this nerve depends on the oxygenation of the blood we are again brought to regard the constitution of the blood and its stagnation as the original causes. And this is indicated also by the sleeplessness and shivering which invariably accompany deranged circulation and which precede the disturbance of the digestion. Such derangements of the circulation may doubtless be in part due to atmospheric influences, so that we must also regard small-pox as depending on certain states of the weather—a fact which is quite patent in the case of black-pox in which decomposition of the blood is involved.

In every case of small-pox insufficient electricity in the nerves and insufficient oxygen in the blood assist as contributing causes of the infectious chemical disintegration of the lymph, and the curative treatment must above all take these things into consideration. What treatment should be applied will occupy us in the fourth part.

CONSUMPTION.

It is but natural that consumption—a disease which as shown by statistics' is responsible for one third of the death rate—should awaken a most general interest. The number of treatises on this subject is incalculable and the confusion prevailing among medical practitioners is in the case of no other disease so great. This is due to the fact that hitherto only the alterations in the lung tissue which are visible to the eye have been studied, but the chemical processes which form the basis of these changes have been left out of consideration. It happens in this case also that the law deduced from Stenton's experiment comes into play—namely that no part of the body can perform its functions properly which is not regularly supplied with oxygenated blood and from which the blood charged with carbonic acid does not flow away regularly. A deciding proof that this law also holds in the case of consumption, is afforded at once by the fact that the inflammatory condition due to stagnation of the blood occurs first of all in the upper part of the lungs near the collar bone. It is just there that the fine ramification of the pulmonary arteries reaches its extreme degree, and it is plain that the finer and more delicate the arterial ramifications through which the blood flows are, the greater is the tendency, in accordance with physical laws to stagnation, and that it is there where it will first occur. This is still more the case with individuals who have a high thorax, for the blood coming direct from the heart through the pulmonary artery experiences increased difficulty in rising upwards, as with every inch of distance between the heart and the collar bone the opposition of a higher hydrostatic pressure must be overcome. Everything therefore depends upon an active circulation, if the lungs are to remain in good health. In this latter connection the heredity which has been attributed to consumption in reality depends mainly upon the inheritance of a long thin upper body. It may be remarked in passing that the expansion of the chest in youth largely depends upon the supply of proper gelatinous nutriment which allows the ribs and the diaphragm the necessary substances for their growth both in length and in breadth. Such kinds of nutriment which may be transmuted into blood gelatin are the gluten of wheat flour, the casein of milk, the albumen of eggs and the gelatine extracted from the cartilage of calves and from broken beef-bones.

From this point of view heredity as a cause of consumption disappears. And so far as grief and trouble are concerned, this is to be thought of rather as producing inadequate nourishment and suppressed respiration, which, indeed, act as causes disposing to this disease. In depression of mind the inadequate consumption of food is in part caused by the blood's flowing mainly to the brain, to the neglect of the abdomen, where in consequence of insufficient supply of blood the digestive juices secreted (gastric juice, intestinal juice, pancreatic secretion and bile) are without any strength.

To what extent consumption in prisons is due to bad air, inadequate nourishment and moody states of mind cannot be computed by statistics; but we must regard as main sources of consumption *imprisonment*, confinement in the room at the sewing table or at the writing table for the purpose of procuring a livelihood; thus also here *inadequate respiration* due to insufficient movement of the diaphragm, and *insufficient nutriment*; all kinds of consumption in the end depend on these two causes.

The establishment of a rational *theory* of consumption is, it is obvious, the necessary basis for its therapeutic treatment. Taking up this problem, we shall perceive what it has been customary to regard as the very essence of the disease, was made up of nothing else than separate symptoms. Amongst these may be especially mentioned the *tubercles* from which a special kind of consumption is specifically described as *tuberculosis*, then also the *spitting of blood* and *loss of flesh* (*consumptio*).

In order that my *theory* of consumption or rather to use the right name, the *chemistry of consumption* may be understood by professional men, I think it necessary to put before our eyes the principal anatomical conditions.

The tissue of the healthy lungs is soft, sponge-like, and yielding to the pressure of the fingers, emitting when so treated a peculiar crepitating sound. After such pressure has been removed the lungs resume their original volume, owing to their natural elasticity. This *elasticity* of the lungs is very noticeable; it enables the lungs on inspiration to take up more space and to contract again on expiration—movements in which it accompanies the contraction and expansion of the thorax, which in turn is put in motion by the strong sinewy diaphragm which forms the base and closure of the thorax.

It is plain that the elastic expansion and contraction of the lungs can only take place normally so long as vigorous automatically acting elastic tissue of which *gelatine* forms the basis, is present; while the elasticity of the lungs is diminished when the gelatine, which like all other bodily substances is subject to consumption and change is not renewed by nutrition. Accordingly we find the lungs of a consumptive patient deficient in elastic gelatine, lumpy and contracted, instead of being expanded like a sponge, and when pressed with the finger it does. not give rise to the crepitating sound characteristic of healthy tissue.

It must also be pointed out that in consumption all other organs of which tendinous fibres i. e. gelatinous tissue forms an essential part (spleen kidneys, intestines, peritoneum, cartilage and bones), suffer a decrease in the amount of elastic gelatine, a fact which can excite no surprise when we understand that gelatine as a connected cementing mass forms the basis of our whole bodily structure. It is therefore meaningless to attempt to diagnose a particular kind of "tuberculosis" according to the organs attacked.

Owing to the due amount of gelatine the tough elasticity of healthy lung-tissue is such, that it is extremely difficult to cause it to tear by forcibly blowing it up. The case is very different in advanced consumption, when the amount of elastic tissue has in consequence decreased. The elastic tissue originates from the ramifications of the bronchi which are connected with the cartilaginous larynx, and the structure of which is interspersed with cartilaginous rings. This cartilaginous substance imbedded at certain intervals in the texture of the bronchial ramifications otherwise formed of smooth muscular cells only disappears when the bronchial ramifications have reached a fineness of 1 millimeter. From this point onwards the walls of the air passages only consist of smooth muscular tissue.

In order to know the chemical nature of the tough mucus (sputum) ejected from the air passages of consumptives we must also consider the skin which lines these passages of the lungs, and the blood vessels connected with them.

This covering membrane extending from the vocal chords to the finest bronchial ramifications is covered with ciliated epithelial cells, which are to be regarded as the terminal shoots of the pulmonary nerve. This epithelial lining membrane gives place in certain regions, namely where the ramifications decrease in diameter to three tenths of a millimeter, to ordinary nerveless flat-celled epithelium formed of gelatine, which is prolonged to the bullular alveoli into which the fine bronchial ramifications enlarge. The inner surface of these alveoli is traversed by capillary blood-vessels, and the alveoli themselves are separated or we may say connected—by elastic, gelatinous connective tissue.

This anatomical presentation shows that capillary blood-vessels, gelatinous connective tissue, and nerve substance in the form of ciliated epithelium are, in the region of the alveoli, found close together. While now the tubercles are situated in the parenchyma which is formed by the connective-tissue surrounding the alveoli, the expectorations on the contrary originate in the lung-passages.

These two products of decomposition (tubercles and sputa) though oppositely complementary to one another, must be considered separately if their chemical constitution is to be understood. We will therefore first of all proceed to describe the peculiar chemical decomposition of the gelatinous connective tissue, to which the name of tuberculosis has been given.

It is well known that all protoplasm contains sugar. In many cases it may be proved to be present as such, in other cases as a glucoside, i. e. in a dehydrated form. In still other cases only peculiar decomposition products are recognizable, so to speak chemical fragments of the sugar molecule, as also combinations of such sugar molecules with ammonia. We have already spoken of that very common product of decomposition of sugar called lactic acid and we will now return to the consideration of it. Lactic acid has quite different properties from e. g. acetic acid. The reason is to be found in the chemical peculiarities of the two acids. While in lactic acid (COO, CHH, CHH, HHO) 1 carbonic acid, 1 water, and 2 hydrocarbons are together, the latter conferring the so-called aldehyde or reducing character to the compound, acetic acid (CHH, CHH, OO) presents quite opposite characteristics since 2 hydrocarbons which it also contains as well as the lactic acid, are saturated by two atoms of oxygen.

These 2 atoms of oxygen confer a positive electrifying power on the acetic acid which power as we know holds the chemical constituents together. In opposition to this, lactic acid which is a product of fermentation absorbs electricity with the result that it disposes other material to a similar chemical decomposition

If 2 atoms of oxygen should be combined with a molecule of lactic acid and 1 of carbonic acid, and 1 of water be separated from it, it would be thereby converted into acetic acid. Such being the chemistry of the question, it is plain that acetic acid is distinguished from lactic acid, by the fact that the former is saturated with oxygen, and the latter is hungering for more.

This hunger for oxygen accompanying the formation of lactic acid plays an important rôle in the disease called tuberculosis, the signification of which we will proceed to study more closely after having introduced certain substances which occur in our organism as derivatives of lactic acid and so of sugar.

1. *Sarcosin* = CHH, CHH, COO, NHHH. This is apparently lactic acid in which ammonia has replaced the water; but it would be more correct to regard it as sugar, which by combination with 2 ammonia and separation of 2 water particles has become 2 sarcosin. This substance is not found in an isolated form, but it occurs in combination with urea as the basis of flesh, *creatin*, and with dehydrated sugar as hippuric acid.

2. *Hippuric acid* = $C_6H_4O_2$ + CHH, CHH, COO, NHHH, minus H_2O. This substance is evidently a combination of 3 molecules of sugar with 2 of ammonia, with separation of 5 molecules of water, and division into 2 equivalents of Hippuric acid.

3. *Tyrosin* = $C_9H_{11}NO_3$. According to its chemical formula, tyrosin may be regarded as the basis of Hippuric acid ($C_9H_9NO_3$) in so far as the latter is formed from tyrosin by the addition of 1 atom of oxygen and the separation of 1 molecule of water. According to its origin it is a product of the oxidation of gelatine based on olein and ammonium carbonate, namely a combination of C_6H_{12} with 2 (NH_3 CO_2). The material of sarcosin is already recognizable in such a group, and sarcosin is itself contained in tyrosin. The possibility of grouping it together to form leucin (C_5H_{10}, CO_2, NH_3) is also apparent, as also that the three substances leucin, tyrosin, and urea are oxidation products of gelatine resulting from the combination of 2 parts of gelatine with 2 of oxygen.

It is, however, only to be regarded as a short way of expression, when gelatine is represented as based on olein and ammonium carbonate. I ought properly to say olein, carbonic acid, and ammonia, which amounts to ammonia and dehydrated sugar, so that the tasteless gelatine resulting from the combination of sugar and ammonia is most closely related to gum arabic which consists of sugar and lime.

Under such circumstances namely since sugar is the basis of gelatine, the occurrence of lactic acid as a product of decomposition of the gelatinous connective tissue has nothing surprising about it. And when lactic acid has once been formed in the tissue as a result of imperfect respiration, a second reason for the formation of tyrosin is then supplied, for the lactic acid combines with the ammonia of the leucin to form sarcosin, while the rest of the leucin ($C_6H_{10}O_2$) imperfectly oxidized, after separating water and in the form C_6H_4O combines with the sarcosin to form tyrosin. That this is the case I have proved chemically by allowing commercial gelatine dissolved in concentrated lactic acid to stand for some months. One day the characteristic "tubercles" crystallized together of tyrosin, were found at the bottom of the glass jar.

The occurrence of tyrosin in anatomical preparations of flesh preserved for years in diluted alcohol is to be attributed to a similar process. The weak alcohol prevents a rapid oxidation of the flesh and the restricted access of oxygen causes the lactic acid to split off, producing tyrosin.

A similar process also takes place in the stomach, owing to the formic acid contained in the gastric juice which possessing the character of the aldehydes which consume oxygen, and assisted by the temperature of the blood in a very short time causes the albuminous substances present to disintegrate into leucin, tyrosin and urea.

Everywhere throughout the body where gelatine exists (i. e. everywhere) this chemical disintegration of the gelatine giving rise to tyrosin (forming tubercles) may take place whenever the abundant supply of oxygen is lacking. It is on this account that young persons readily fall a prey to consumption if they are confined to heated rooms and have but little exercise, even like as the apes in our zoological gardens. A similar fate befalls the Eskimos who are brought to our climates to be exhibited, and instead of the cool concentrated polar air to which they are accustomed, are confined in this heated air which is still farther vitiated by the presence of many spectators.

A similar effect is produced on weak constitutions by smoking tobacco. As the nicotine of tobacco-smoke contains ammonium cyanide and this combines with the hemoglobin of the blood, this becomes unable to combine with the respired oxygen, and so from a deficient oxidation of the substances of the body there follows the formation of tyrosin in parts of the body where it certainly can do no *good*. As such it is not, indeed, productive of any direct *danger*. For just as the tyrosin and leucin produced in the stomach may combine to form normal tissue-substance, so the tyrosin formed from the substance of our body may undergo a chemical conversion, in the case of consumptives the respiration of oxygen produces the most natural change of the tyrosin into hippuric acid which combining with soda or lime passes off from the body through the kidneys so that the cheesy abnormal growths may be caused to disappear completely leaving only harmless holes (*cavernae*) behind them.

This chemical explanation completely deprives the long obscure origin of the "tubercles" of the mysterious and terrifying element. We have made an important step forwards in recognizing in the tubercles i. e. the spherical aggregations which occur in the lungs of the consumptive, tyrosin, and in further discovering their origin to be the combination of lactic acid with leucin imperfectly oxidized, owing to insufficient respiration.

We will now proceed to examine into the chemical constitution of the tough expectorations (sputa). I will preface that nerve-fat (lecitin), when boiled with an alkaline substance (e. g. water of baryta) gives rise to various products of decomposition, namely the so-called glycerine-phosphoric acid—a combination of propionic and phosphoric acids, further stearic acid $C_{18}H_{36}O_2$, oleic acid $C_{18}H_{34}O_2$, palmitic acid $C_{16}H_{32}O_2$, as also the so-called neurin $C_5H_{15}NO_2$.

Neurin according to its formula is evidently produced from the constituent of gelatine, *leucin* $C_6H_{13}NO_2$ which after oxidation by two atoms of oxygen, detached 1 carbonic acid CO_2 and has drawn to itself 2 hydrogen from the stearic acid, thus reducing this to oleic acid.

The possibility of the origin of neurin, however, is in no way limited to its derivation from nerve-fat, for it may also be produced artificially from a chemical combination of trimethylamine (CHH, CHH, CHH, NHHH), ethylene oxide (CHH, CHH, O) and water HHO. From the latter fact it becomes apparent, that all that is required to produce neurin is the combination of ammonia with 5 hydrocarbons and a little oxygen. These 5 hydrocarbon molecules together with ammonia as the essential base of neurin are present in leucin. But the neurin might be just as well produced from ammonia in combination with propionic acid ($C_3H_6O_2$) and two hydrocarbon molecules from the stearic acid of the nerve-fat $C_{18}H_{36}O_2$ which is thereupon reduced to palmitic acid, $C_{16}H_{32}O_2$. Accordingly, if only sufficient ammonia be present, an equally productive source of neurin exists in the fatty material of the nerve-fat as in the leucin of the gelatine. And this is a matter of especial importance, since a poisonous combination of neurin and ammonia is present in almost inexhaustible quantities in the expectorations of consumptives and makes its origin from fat plainly known by its fatty nature which prevents its dissolving in water, while its gluey nature points to its connection with gelatine.

The ammonia required for producing the masses of neurin from fat is obtained from the blood which stagnates in the fine capillaries which traverse the epithelium of the air passages. All stagnating blood whether in or out of the veins produces ammonia, and of ammonia we know, that even in dilutions showing merely a trace, it paralyzes the nerve functions. We must, therefore, be all the more careful not to permit a stagnation of the blood in the veins of the lungs to last any time.

What then is the poisonous substance which originates in neurin and is found in the sputa of those suffering from consumption and in the ill-smelling, putrid expectorations of those afflicted with bronchitis putrida?

This substance is the so-called ptomaine or poison of putrescence (*cadaverin, putrescin*) and may be extracted from the heart as well as from the lungs and liver of corpses, perhaps because all these organs are traversed by nerve substance and blood vessels as well as by gelatinous connective tissue in close juxtaposition.

The chemical formula of ptomaine or cadaverin is $C_5H_{14}N_2$, a composition which shows it to be a dehydrated product of the addition of neurin and ammonia, thus;—

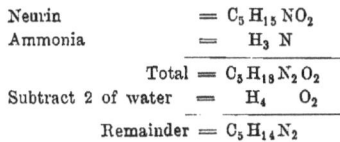

$$
\begin{array}{rcl}
\text{Neurin} & = & C_5H_{15}NO_2 \\
\text{Ammonia} & = & H_3 N \\
\hline
\text{Total} & = & C_5H_{18}N_2O_2 \\
\text{Subtract 2 of water} & = & H_4 \quad O_2 \\
\hline
\text{Remainder} & = & C_5H_{14}N_2
\end{array}
$$

Ammonia has ever a hydrolitic or dehydrating action. From this point of view the oedema or the "hydrops pulmonum" which accompanies consumption is merly a *symptom*; the *essence* of the matter is the appearance of ammonia from stagnating blood.

I would here recall the fact, that the poison of hemlock (coniin) can be prepared artificially from gaseous ammonia, acetone, and ethyl alcohol, while in nature it is produced from vegetable albumen by imperfect oxidation and hydrolysis (detachment of water). And it will I think prove of particular interest, that also certain ptomaines (putrescin) which can also be extracted from putrefying horseflesh, putrefying muscles, fish, and other putrescent substances, especially also from putrescent urine which in cystinurea dissolves the gelatine of the bladder, can also be produced artificially as pentamethylene diamine i. e. as a compound of 2 ammonia with 5 hydrocarbons, with partial oxidation and hydrolysis. Thus on one side ammonia and on the other 5 hydrocarbons from whatever origin, can form the basis of the ptomaines, the presence of which in the expectorations of the consumption presents the proper fatal cause, which spreads the putrescent decomposition to neighbouring tissues so that at a certain stage the walls of a larger artery burst and the blood gushing forth in a strong stream fills the lungs, causing suffocation. The hemorrhages from the lungs at the commencement of the disease are not to be compared with such a fatal flow of blood. The former are more of the nature of a curative effort; for in this case the carbonic acid collected in the blood bursting through the pores mechanically forces the blood from the capillaries, in the same way as the carbonic acid carries the seltzer water with it when the cork is removed from the bottle. This does not show a destruction of the connective tissue in the alveoli, and it is a matter of experience that after such a discharge of blood surcharged with carbonic acid, the health of such patients generally improves, if strength giving nutriments and a suitable manner of living co-operate to the cure.

On the other hand it must be specially emphasized, that a cure can no longer be expected when the destruction of the substance of the lungs has reached a certain point. If, nevertheless, a cure be attempted, a poisonous ammoniacal product of decomposition of gelatine like Koch's lymph can in no way be used. The cases of hastened death that have followed this treatment were much more numerous than has been openly admitted; four were reported to me alone. On the contrary we have in the ptomaine substances which show us the way that a curative treatment ought to pursue; for the ptomaines on treatment with acids lose their poisonous properties. Naturally enough! The poison is just the ammonia; by acids this body is chemically bound and its infectious character destroyed. Is it not then suggested by the nature of the case, that the infecting power of the ptomaines should be destroyed by the periodical inhalation of the vapors of weak, heated vinegar? The other points in the therapeutic treatment of tuberculosis will be dealt with in the special therapeutic part.

SPASMODIC DISEASES.

The first stage on the road towards the comprehension of the causes of disease is due to the English physiologist Stenson. This investigator made a ligature on the femoral artery of a dog and found as a result that the leg became paralyzed and remained paralyzed even after the ligature had been removed. From this it became plain that a system of nerves ceases to fulfil its proper function when its supply of oxygenated blood is interrupted.

This proposition based on experiment was so to say incorporated in the codex of physiology while no practical application has been made of it hitherto. Now, however, the time has come for this instructive experiment to spread its light over a large number of domains that have hitherto remained obscure. It is quite natural to draw the following conclusions from the experiment.

1) The *absolute* withdrawal of arterial i. e. oxygenated blood from any nerve region suspends its functions *in toto*.

2) An *incomplete* supply of oxygenated blood will cause an *imperfect* action of the nervous region in question.

3) If we succeed in supplying a nervous region which is acting imperfectly with an adequate supply of fresh oxygenated blood, the nervous action will be restored.

In pathology a continuous absolute lack of arterial blood such as is produced by a ligature of the tibial artery, scarcely requires consideration; the most that need be considered at least at the commencement of an illness is merely an imperfect supply of oxygenated blood.

Such disturbances in the circulation cause the affection which by many persons is still regarded as possessing something peculiarly mysterious, namely epilepsy, the peculiar nature of which disease I have been enabled to deprive of its veil of mystery, by applying to the problem the conclusions derived from Stenson's experiments, with the result that this disease is henceforth to be included in the list of curable diseases.

St. Vitus's dance and epilepsy which develope among school-children, may often enough be attributed to a defective condition of the blood, and to the weakness of the nerve substance which results from insufficient nutrition. The nerve material also has not sufficient attractive power owing to deficiency in salts of phosphorus—the source of its electric excitability, and of its capacity to attract the blood to itself. The blood also does not contain a sufficient amount of iron to give rise to a more intense attractability; on the other hand it is deficient in lime and sulphur which are required together with the iron for the formation of the red corpuscles which ensure adequate circulation of the blood; and finally it is often deficient in salt particles which electrically excite the walls of the blood vessels and the nerve trunks.

In such a state caused by inadequate nutrition, interruptions of the circulation accompanied by electroscopic manifestations in the form of twitches and convulsions, are easily explicable.

The first stages of such a spasmodic condition with children is shown by their inattention and absent-mindedness. These signs are almost universally regarded as deserving of punishment and they receive it; but often enough the children are not to be blamed, these states being merely the results of insufficient nutrition and the monotonous methods of teaching.

After what has been said on pp. 46 and 47 about exhaustion of the nerves, it is sufficiently clear, that where but little nerve-material exists, but little can be carried away, or set in action. Who can count those who dream or even sleep during many a sermon! And from undeveloped children is expected what can not be accomplished even by adults namely to resist the tendency of eye and ear to become wearied by uniform monotony?

The cramp-like rigidity of persons who have gone to sleep from nervous exhaustion, mostly anaemie individuals (those hypnotized)— throws considerable light an spasmodic states in general.

If we once realize that our nerve-substance as a consequence of the salts of phosphorus it contains, possesses its own electricity in virtue of which it exercises an attraction on the blood, we understand also how such electricity in accordance with the nature of the electric fluid, tends to extend itself over the largest possible area, i. e. to separate as far from each other as possible i. e. into halves, and thus to flow to two terminal poles causing under certain circumstances even radiation; when for instance the nervous system of the intestines and the brain owing to inadequate circulation of the blood cease to maintain their connection, or when through a check in the circulation caused by the accumulation of carbonic acid which cannot support the process of combustion, the nerve substance ceases altogether to be consumed. In the latter case the constant escape or radiation of the electricity accumulated at the terminations of the nerves is identical with the continuous electric discharges in the muscles producing tetanus, i. e. convulsive twitchings which follow one another so rapidly that no interval is perceptible between them, whereas when the interruptions of the circulation are only temporary those twitchings or convulsions occur, which are known as St. Vitus's dance.

Tetanus shows that the electric fluid in its attempt to reach the terminal poles by the shortest route tends to stretch out the nerve trunks into a straight line and they draw with them the substance of the muscles in which the substance of the nerves is imbedded.

Thus the *distortion of the muscles of the face* is due to the stretching of the third and the seventh pair of nerves (the *motores oculi*, the *pathetic*, the *trifacial*, and the facial nerves). The checked respiration indicates that the tenth nerve-pair (pneumogastric) is in spasms and the foam at the mouth points to the paralyzed state of the nerves of the salivary glands. Loss of consciousness shows that the blood has withdrawn from the brain into the veins of the abdomen as is also shown by the paleness of the face.

Taken altogether, the deduction from Stenson's experiment, which shows that no nervous region can properly perform its function unless it be supplied continuously with oxygenated blood, applies also to spasmodic diseases. In this connection I may mention a case of epilepsy which occurred from the time that the patient had fallen on the ice on the back of the head; since at the same time the patient began to suffer from diabetes, there could not be much doubt that the various symptoms were due to an injury of the pneumo-gastric nerve which originates in the medulla oblongata, and disorganization of which causes inadequate oxidation of the blood which is ever the cause of spasmodic conditions in general as well as of epilepsy in particular.

Yawning—is a cramp of the tri-facial nerve due to exhaustion of the brain through the accumulation of carbonic acid after the consumption of the nerve oil in certain parts of the brain.

Cramp of the jaws in new-born children may also be attributed to a deficiency of oxygenated blood owing to its being impoverished as to iron, lime, sulphur and salt particles by feeding with diluted cow's milk. The death which often results is not the result of the cramps, but of the paralysis of the nervous system due to a general want of oxygenated blood.

The cramps in the calves which come from crossing one's legs, and is due to the compression of the popliteal artery which behind the knee approaches close to the skin, reminds us of Stensons experiment.

Of a similar nature is the piece of legerdemain which Moses learned of the Priests of Sais and imparted to his brother Aaron, of throwing a snake into tetanus so that it may become rigid like a staff by strongly compressing the artery of the tail. By this means the end of the spinal column is cut of from a proper supply of blood, and the whole spinal marrow experiences a longitudinal rigor since its proper electricity flows toward the two terminal poles.

Epilepsy is often connected with interrupted catamenia. Here the deficiency in the blood of nerve-quickening oxygen is manifest. It is the same when epilepsy is connected with sterility and cramp of the uterus.

The crawling sensation as if from a number of ants which precede an epileptic attack points to the gradual withdrawal of the electric induction current which under normal conditions is active in the walls of the blood-vessels owing to the circulation of the iron-holding blood. The patients first notice this withdrawal from the feet toward the heart the principal blood receptacle. The second step is that the blood withdraws from the brain and unconsciousness ensues. Finally all the blood collects in the capacious blood vessels of the intestines, convulsions and rigidity then set in as the physical expression on the one hand of the interruption of the electric current in the walls of the bloodvessels, and on the other hand of the continuous discharges from terminations of the nerves.

If after all these considerations, it can no longer be a matter of doubt that disturbances in the circulation give rise to epilepsy, the treatment of this complaint must consist in establishing the conditions under which a regular circulation of the blood can take place. This we will consider in the therapeutic part of our work.

IV

THERAPEUTIC PART,

OR

CURATIVE TREATMENT.

In the pathological section we have described certain characteristic affections which plainly manifest themselves as disintegrations of the albumen of the blood, of the lymph and of the gelatine, or as disturbed nervous function, and in all the various cases we have shown that the diminution of the electric fluid is the real cause leading to disease. Since the albumen of the blood, gelatine, lymph, and nerve-substance have chemically much in common, and physiologically as well as anatomically pass one into the other, it is consequently plain that there must be states of disease in which all the different regions of the body will participate, though in different degrees; and since in consequence the appearance of the disease is subject to considerable variation it becomes explicable, that for thousands of years the mistake has been made of supposing that there are as many different diseases as there are modifications in the nature of the disease.

We frequently hear the remark—"I cannot yet say what disease will here develope." In consequense much time is lost which might have been employed in arresting the progress of the chemical decomposition which has set in. For it is plain, that chemical action once begun will not stop of itself, but will extend its sphere of operation, if sufficient time is given. This fact that also a chemical process requires time, is of particular importance for therapeutics. The important point is to act quickly enough to prevent chemical decomposition in no matter what sphere. For this purpose there are two universal medicines which make impossible the spread of most of the acute disease—namely Vinegar, and Glauber's salts. Rubbing the whole of the body with ordinary table-vinegar bestows fresh electricity on the skin and to the terminations of the nerves; and a solution of 1/6 oz. of Glauber's salts and 1/6 oz of table salt in 1 litre (1.05 quarts) of water used as a beverage will arrest the chemical decompositions attending the so-called inflammatory complaints, and as all salts act as electrical excitants, the electrical fluid, instead of disappearing, will be materially increased. The chemical explanation of this fact is very simple and is as follows.

In the basis of muscle-substance, creatin, is contained intermolecularly the body sarcosin or hydrocarbonate-gelatine-sugar; the same is the case with leucin and tyrosin, from which chyle and thence the lymph are formed, and of course also with the gelatine of the blood. Now glycocoll or gelatine-sugar (COO, CHH, NHHH) possesses the faculty of uniting chemically with salts, e. g. Glauber's salt (sulphate of soda), the soda of the latter attaching itself to the carbonic acid of the glycocoll and the sulphuric acid to its ammonia. In this manner the glycocoll is doubly anchored, and the group containing glycocoll is consequently rendered safe from decomposition.

Still better than Glauber's salts and common salts is a mixture of salts exactly corresponding to the composition of the salty portion of the blood, and containing besides Glauber's salt and ordinary salt, sulphate of potash and phosphate of soda. All these salts combine in their tendency to form combinations with the glycocoll of the different parts of the body, as also with the urea which exists *in posse* in the bases creatin, sarcin, xanthin &c. By such natural salts of the blood the flesh substance is preserved from decay and putrefaction, i. e. from chemical disintegration.

A mixture of salts formed in faithful imitation of the salty contents of the blood I term *physiological normal salt*)* and a solution of ¼ ounce of it in 1 quart of water I simply call *physiological salt water.*

By drinking this "physiological salt-water" the most natural means of transfusing blood is provided, as it corresponds in composition to the serum of the blood and is absorbed in the intestines by the lymphatics whence it passes through the lymph to the blood. The lymph being at once electrified and rendered capable of resisting chemical decomposition, the whole organism is beneficially affected, for the electric "fluid" passing along the walls of the chyle-vessels passes into the whole system, the result being that the effect on the patient is immediately visible. This weak solution of salts does not introduce anything of a foreign nature into the body, but only adds substances which are absolutely indispensible, as at least ¼ oz of salts are passed off by the urine every 24 hours as a consequence of the oxidation caused by breathing, and must be restored to the system, else the amount of bodily electricity will be diminished. It consequently forms a beverage suitable for every condition.

---

*) I have communicated to Messrs. Boericke & Tafel of Philadelphia the exact composition of this salt for the United States and Canada.

When we recognize that most of the so-called diseases are either only varieties of those already described or else mere disconnected symptoms, we find that general therapeutics resolves itself in all cases into the endeavor to set at work those agents which produce electricity, which either separately or combined are capable of producing the desired effect in the case of the special disease under consideration.

Having made these prefatory remarks, it appears to me most practical in discussing the affections known by special names to treat pathology and therapeutics as belonging together.

Asthma. Asthma is in German also called thoracic spasm (Brustkrampf), since this affection is characterized by a spasmodic state. Though the nerves without doubt are the agents in this disease, the accompanying symptoms (hemorrhoidal discharges, gout, rheumatism, itch &c) show that the cause is to be found in the defective constitution of the blood. Owing to such defective constitution of the blood there arise disturbances in that part of the circulation passing from the right ventricle of the heart through the pulmonary arteries, and the lung capillaries and the pulmonary veins, back to the left auricle. Asthma is due to disturbance in the circulation in this part of the vascular system, and it affects those portions of the pneumogastric nerve which end in the air cells. The result is to cause an interruption in the action of the cells until sufficient blood flows in from the general vascular system to restore the circuit and so remove the cramp. As attacks of asthma usually take place at night when most of the blood is concentrated in the viscera and intestines for the sake of the new formation of blood, it is plain that the cause of the cramp of the cells is similar in character to that of epilepsy, (which is due to scarcity of blood in the brain,) namely temporary local scarcity of blood in the pulmonary regions. A spasm always manifests itself when the circulation stagnates and thence the proper electricity of the nerves comes into play. In this connection it should be remembered that both convulsions and spasms can be produced artificially by the passage of an electric current through a muscle. Whenever the current is made or broken there follows a convulsive contraction of the muscle; if discharges are allowed to pass at intervals immeasurably short, then tetanus of the muscle is produced owing to the fact that electricity follows the *straight line* as the shortest course. The latter fact may be well observed in the case of foxes that have been poisoned by strychnine. We there find that in consequence of the tetanus, the head is bent into the thorax, so that brain, medulla, and tail form one straight line. In this case the whole nervous system is affected. In other cases in which lungs, stomach or uterus are affected the spasmodic manifestation is confined to some special region of the body; but in all cases the activity of the nerve-electricity which is based on salts of phosphorus, is to be traced to *disturbances in the circulation*.

In the case of asthma the fine branches of the pneumo gastric and their plexuses with strands of the sympatheticus participate in this tetanic condition. In correspondence with this state of affairs the last fine air passages in whose annular muscular fibres these nerve fibrils are imbedded, are similarly affected whenever they experience an interruption in the regular supply of arterial blood. In this the central blood reservoir—the heart—takes a leading part, so that not unfrequently emotional causes give rise to attacks of asthma. A similar effect is produced by sexual excesses which naturally must be accompanied by disturbances of the circulation. Similarly attacks of asthma may be provoked by indigestion causing a long continued concentration of blood in the abdominal regions.

It thus appears that a tendency to asthma is not to be considered a special disease, but merely a symptom, and various circumstances show that the real cause is an inadequate supply of red blood-corpuscles. This is shown by the rheumatism which so frequently accompanies asthma and which also does not present a disease but only a symptom. When we proceed to enquire to what this deficiency of red corpuscles is due, we find that those blood corpuscles which are used up in respiration cannot be replaced, unless the nutriment supplied to the body contains sufficient iron, lime, and sulphur to adequately replace the amounts of these substances which daily pass away in the urine. For lime and sulphur are indispensible for the formation of normal albumen of the blood. This lack of lime and sulphur is also indicated by the occurrence of itch which frequently accompanies asthma. The itch parasites are produced from the phosphatic material of the epithelium full of nerve-fibrils, when the blood is deficient in lime and sulphur. It was from this reason that the first Napoleon suffered from itch, as his principal daily food insisted for a long time of the brain of an ox. The brain substance contains neither lime nor sulphur. This extraordinary diet gives us the key to many strange pathological problems, notably the extreme nervousness of Napoleon which sometimes reached to the verge of madness.*)

While thus speaking of the specific disease of asthma we are in the very heart of universal pathology; so intimately the single point is connected with the whole. It is here as with the hundred gates of Thebes; no matter through which gate a man entered he came in every case to the market place in the centre.

When once we know that normal blood contains twice as large a proportion of sulphates as of phosphates, we understand that any

---

*) What else was it but insanity that caused Napoleon in Russia to command regiments to take their position of which he knew that only the commanding officer was alive, and that when this was pointed out to him he brusquely declared that he did not desire to be reminded of the fact! (see Ségur).

considerable variation from this proportion must necessarily produce corresponding disturbances. Sulphur keeps the balance against phosphorus as I have explained at length in my book "Das Leben". The irregular behavior of the phosphatic nerve-substance is kept in bounds by the due amount of sulphur which the blood contains. Therapeutics thence concludes, that nervous irritability may be cured by supplying sulphur to the blood and lymph. But not only nervousness but many other symptoms also enable us to understand the factors which are really at work here. Many subjects, we repeat it, appear completely incomprehensible when considered by themselves; but when brought into juxtaposition with others they all gain clearness and comprehensibleness. By this method of viewing diseased conditions apparently distinct, from a common point of view, pathology and therapeutics become simplified to an astonishing degree. The association of asthma with obesity enables us to take a further step along the same road. This also clearly shows the deficiency of red corpuscles, for if enough of them were present, sufficient oxygen would be respired and the fat in consequence would be oxidized instead of accumulating (cfr. p. 76).

Piles and Asthma are also frequently found in company, and both must be claimed as being due to insufficient oxygen in the blood. In the case of piles the blood is overcharged with carbonic acid; this latter forces its way in gaseous form through the capillaries and drives the venous blood before it. If, however, through preparations of lime, sulphur, and iron more red corpuscles are formed the circulation regulates itself and, indeed, in consequence of the increased respiration of oxygen.

We have thus a complete cycle of pathological conditions which according to the teaching hitherto prevalent were regarded as totally heterogeneous and disconnected, namely *obesity, nervousness, rheumatism, piles, itch* and *asthma;* these must not be regarded as independent diseases, but merely as the symptoms of the real cause of disease, an insufficient number of red blood corpuscles due to an inadequate supply of sulphur, lime, and iron, the result of imperfect nutrition. As a proof of this we will give a concrete case:—

A lady writes:—

S., Feb. 27. 1891.

My husband suffers from weak nerves and is much inclined to corpulency. His weight has increased this winter by more than 13 lbs. and his general health has grown considerably worse. Especially it is sleeplessness of which he had to complain for weeks. His night's sleep often lasts only 3 or 4 hours after which he cannot find any refreshing rest. His disposition to asthma also seems to increase and is often the cause of insomnia. Please send me the suitable remedies for curing this disease.

R. A.*)

The above letter came from Switzerland where wine drinking is general. Wine contains neither lime, sulphur, nor iron to any appreci-

---

*) All letters printed in this treatise are in the hands of the publishers.

able amount, for during fermentation the lime is precipitated as tartrate of lime and the iron as tartate of iron, and the sulphur contents of grapes is only 1/20th of the weight of their ashes. Wine drinking, however, causes a considerable amount of urination, which carries away pretty constant amounts of iron, sulphates, and calcium compounds from the blood. It follows, therefore, that when wine is drunk regularly every day, a condition must in time necessarily arrive, when the blood contains too little of these substances to enable it to form a proper new supply of red corpuscles. It is then that asthma, corpulency, and rheumatism commence their rule.

Now it is a matter of experience that asthma and corpulency may be cured by copious consumption of the Driburg mineral water, the analysis of which shows that it contains per quart 15 grains of sulphates of soda and magnesia and the same amounts of sulphate and even more of carbonate of lime together with over 1/2 grain of carbonate of iron. Lime, iron, and sulphur, I repeat, are the necessary basis for the formation of ordinary normal albumen of the blood, the capacity of which for absorbing oxygen cures asthma. Iron and sulphur also remove nervous irritation (insomnia) because the sulphur of the blood keeps the phosphorus of the nerves in equilibrium. The extraordinary efficacy of sulphates in connection with iron and lime in cases of corpulency is shown by the effect of the Ferdinand spring at Marienbad, which contains 1/6 of an ounce of sulphate of soda, nearly 1 grain of sulphate of potash, about 15 grains of carbonate of lime together with magnesia and nearly 1 grain of ferrous carbonate pro litre. The Rakoczy spring at Kissingen has about half the strength, but is of a similar constitution with the exception of the iron.

What is effective in all these three springs is iron, sulphur, lime, and magnesia.

In nature magnesia accompanies lime almost everywhere. For instance our bones contain in addition to phosphate of lime a certain quantity of phosphate of magnesia, and magnesia, also takes part as well as lime, iron, and sulphur in the formation of the albumen of the blood. This circumstance we have to bear in mind in treating asthma, corpulency, rheumatism, piles etc. without the assistance of the above named springs which can only be made use of in summer, when the skin evaporates larger quantities of moisture and there is less need to consume large quantities of concentrated nutriment to keep up the bodily heat in the abdominal regions.

With this view I have found effective a treatment of asthma with lime, sulphur, and iron, extending over a number of months in the following manner:

1. Morning and afternoon stirred into a cup of black coffee 7 1/2 grains of calcium-magnesium-phosphate on the chemical proportions corresponding to these of the albumen of the blood and of the bones.

2. At noon immediately before dinner 3 to 4 grains of hematite iron with about the same quantity of common salt in a spoonful of water. The same effect is produced by the first homoeopathic trituration (1 : 10.) as the number of points of contact with the walls of the stomach and intestines is increased, and the absorption more complete.

3. At night about an hour before retiring 8 grains of flowers of sulphur stirred into a table-spoonful of milk.

4. Every second day 4 grains of amorphous silicic acid (silicea)*) fasting, or its first homoeopathic trituration stirred up with a cup of warm boiled milk to which a pinch of salt and a tea spoonful of sugar have been added. By means of this preparation the constitution of the blood is improved in a direction the importance of which has hitherto been but little appreciated. The ashes remaining from healthy muscle contain at least 2 % of silica. This silica comes from the blood which owing to its warmth easily keeps 1 part of silica per thousand in solution, a portion of which, indeed, passes off with the urine. The blood takes up the silica from the chyle which in turn obtains it from the food, assuming that the latter contains a sufficiency of silica. Unfortunately so far as ordinary wheat flour goes this is by no means the case. In vegetable foods the silica is combined with cellulose and surrounds the grains of cereals, as a membranous covering. Now as this remains behind as bran after grinding and bolting, this source of silica is removed from ordinary white flower, a fact which largely explains the spread of certain modern diseases. Peas and beans which besides do not often constitute a part of the *daily* food, contain only about 3 grains for every two pounds which considering the small amounts of them that we consume is very little. Much more favorable in this respect are the various kinds of millet giving as much as $1/3$ to $1/4$ oz. of silica per pound.

The physiological importance of silica is the following: (1) It makes the muscles firm, for it protects them against chemical decomposition and has consequently an antiseptic action. (2) It warms the blood by isolating and keeping together the electricity imparted by its salty constituents (an electric current is in its effects equivalent to the production of heat). (3) As nerve substance contains approximately the same amount of silica as the albumen in the blood it forms as it were a connecting link between blood and nerves, lastingly preserving the proper relation between them.

In passing it may be remarked that the restorative effect of mineral waters on weakly patients is indubitably to be ascribed to the amount of silica they contain in solution. Throughout the winter we are silica-starved owing to the white wheaten flour we consume and we are therefore obliged to make up for the deficicency in the summer by

*) In Philadelphia from Boericke & Tafel, pharmaceutists.

drinking the mineral water of springs which through their contents of salts earths and silica exercise a curative action. For instance the Marienbad Ferdinand spring contains 1½ grains of silica per quart, the Kissingen Rakoczy 4 grains and the Salzburg upper spring ½ grain.

In passing it may be mentioned that silica may be used with advantage by all persons whose hair is falling out. Hair requires for its growth among other things also silica. If there be a deficiency of silica in the blood the hair is not properly nourished. It loses its connection with the nervous mucous membrane and falls out. But the hair ceases to fall out and fresh hair begins to grow as soon as silica and flowers of sulphur are used.*)

Is it not an interesting fact that asthmatic persons are often "blessed" with a bald patch? Having discovered the relations subsisting between asthma and other forms of disease we also find light thrown on the asthma (Spasm of the Glottis) which attacks children that have been brought up by the bottle, and which frequently ends in death from suffocation. It is easily seen that the practice of diluting cow's milk with water or gruel diminishes the percentage of iron, sulphur, lime and other salts in the milk, and that consequently the capacity of the blood for absorbing oxygen and its proper circulation are influenced prejudicially. Tonic spasms of the glottis in infants is thus allied to tetanus of the jaws and convulsions in teething. The proper preventive and curative treatment consists in giving undiluted milk putting a pinch of salt and a tea spoonful of sugar in each bottle full. Milk is the form of nourishment containing the highest percentage of lime. Of the 108 grains of ash from a quart of milk more than one half consists of compounds of lime and magnesia—the sulphates only forming about one seventieth part of the phosphates, whence milk appears less suited for adults than for children who require a great amount of phosphorus compounds for the formation of nerve substance.

Basedow's Disease. The sum total of symptoms accompanying this complaint shows that it is, beyond the possibility of a doubt, a case of partial paralysis of the sympathetic nervous system. This is plainly shown by the increased rapidity of the pulse. The heart receives its principal impulse from the cerebro-spinal nervous system, namely from the branches of the vagus, while the strands from the thoracic ganglia of the sympathetic with which the vagus fibres are interwoven like meshes exercise a retarding influence analogous to the retarder on the pendulum of a clock. If the latter be removed and the weight still allowed to act, the hands run round at a high speed and only

---

*) The Baroness von L. wrote to me on March 1st 1891. "The new hair is growing; and the little hair I still possess does not continue to fall out as much." The husband of this patient wrote on May 1st 1891: "The hair has grown fabulously, it is already a finger's length and of great density."

regain their proper motion when the pendulum is put back in its place. The same sort of thing occurs in the case of the action of the heart whenever the regulator—the sympathetic system is disturbed. That a relaxation of the fibrils of the sympathetic nerve takes place is shown by the enlargement of the blood vessels which are largely constructed of spiral windings of the sympathetic fibres. With this enlargement is intimately connected the distension of the superior thyroid artery and of the branches of the ophthalmic artery. Both of these arteries are supplied by the external carotid from which (indirectly by means of the internal mammillary) also the infra-orbital receives its blood. Accordingly a swollen neck and protruding eyes are among the symptoms of Basedow's disease. The larynx and bronchi are often affected at the same time by catarrh, also due to the local congestion of blood.

Young girls before reaching puberty often suffer from Basedow's disease. They may be cured by treatment with sulphur, lime and iron, which give rise to the formation of a greater number of red blood corpuscles, so restoring the circulation by a sufficient absorption of oxygen to its normal and proper condition. What paralyzes the circulation and the vascular nervous system is the surcharge of the blood with carbonic acid. In the case of the male sex, it may be cured in the same manner by a methodical use of the preparations of sulphur, lime and iron with the addition of "physiological salt-water" to restore its electrifying power to the blood. I will give the following as an example of a cure, without mentioning numerous cures among the female sex:

W., 27. March 1890.

I am a man of only 41 years. Health generally good, and very good appearance, until in the autumn I got a *thick neck* and in the middle of December influenza with a violent *bronchial catarrh*. Then *swollen feet, trembling in the limbs*, and *intense palpitation*. Dr. Sch. declared it to be Basedow's disease, and recommended the electro-therapeutic institute of Dr. S. Since February 19th I have been going daily to the institute, but I feel only an insignificant alleviation therefrom all this time. An *unnatural appetite**), and I have so many *sleepless nights*.**) Please advise me. Z. V.

I ordered this patient to give up beer and ordered the same medicines as mentioned for asthma. His doctor had been employing the galvanic current from the back of the neck to the goitre and the solar plexus, changing the poles about. After 5 weeks I received the following:

May 11. 1890.

I received an account from my brother-in-law in Vienna who was suffering from Basedow's disease. I strongly advised him to apply by letter to you which he did and he has cause to be satisfied with the result. I took him home with me to Moravia as he had been granted a two months furlough. His state at present is as follows:

---

*) The pneumo-gastric nerve is not affected.

**) Caused by congestion of the blood in the occipital artery which originates from the external carotid artery at an equal height with the external maxillary artery.

His appearance still somewhat indicates suffering, and the eyes are still somewhat prominent and sore and moist till about 10 a. m. His sleep is sound and his appetite enormous. Since 8 days we make daily excursions in the forest, and he is very glad to see that the audible palpitations from which he suffered so much in Vienna have disappeared. He can already climb moderate heights without much trouble. During such excursions he feels a craving for beer, but I have forbidden it. As a postscript to his letter of March 24th I would mention that in his arduous position, he preferred the animal to the vegetable diet and drank about 3 or 4 glasses of beer a day. After receiving your advice he threw all prescriptions aside and in 14 days was glad to perceive an improvement. Before leaving he went to see his Doctor who after percussion and auscultation, looked in his face and asked him "What have you taken?" He replied "what you ordered!" "Fortunate man! replied the Doctor, "in other cases this disease often lasts months and even years. Now you will be quit of it in 2 or 3 months" &c.   E. S.

As a matter of fact complete recovery through methodical improvement of the blood requires on an average a period of 3 months.

Calculus and Gravel. The chemical composition of stone enables us to understand that they are caused by an insufficiency of oxygen and of mineral salts in the blood. That is to say, they consist in the main ot phosphates and oxalates of lime and of compounds of urea.

As far as uric acid is concerned, it is well known that its difficult solubility in the serum of the blood favors its separation, so that thus the foundation is laid for concretions which in the capillaries impede the circulation and thence produce local pains and spasms. This separation of uric acid always proves as cannot be sufficiently repeated, that the constituents of the blood have not been sufficiently oxidized into water, carbonic acid, nitrogen and urea (—urea in contradistinction to uric acid being easily soluble). Now since uric acid forms with mineral salts double salts which are easily soluble, we are led to the conclusion that in patients who suffer from calculus the blood is deficient in mineral salts. The fact that calculus only occurs in the case of persons who indulge in a meat diet and drink spirituous liquors (wine, beer, brandy), breathing at the same time impure air in ill ventilated rooms, harmonizes with this conclusion.

Phosphate of potash is almost the only mineral base present in meat, wine, and beer—merely traces of other salts especially such as the sulphates being present. Since the so-called "organic" material of the muscles is based on uric acid and its compounds, the complete oxidation of muscular substance requires a large amount of oxygen. We notice in the case of carnivorous animals (e. g. wolves, jackals, hyenas, leopards) when confined, and of the dog when at large, a restless movement hither and thither, i. e. a determined desire for motion and respiration. The carnivorous birds too, the raven, vulture and eagle move in cool pure mountain air. Breathing cool air is equivalent to drinking water (see p. 65). This example should be followed by the meat-eating portion of mankind by either breathing cool air or at least drinking an abundance of water for the purpose of converting the prussic acid basis of the urates into formiate of ammonia and so forming a basis for fat and gelatine.

Instead of making use of these health giving factors the meat-eaters spend too great a portion of their time in hot rooms deficient in oxygen, and drink beer or wine to quench their thirst, the alcohol of which for its oxidation withdraws oxygen from the blood, consequently the urea compounds can not be sufficiently oxidized and half burned portions remain (see p. 87). To dissolve and exhale these remnants the patients have to visit the baths, breathing fresh cool forest air and drinking mineral waters, the salts of which make the urates soluble. In addition warm baths help to distent the capillaries which had become stopped up through excreted urates. In this way the blood which has ceased to move in the passages is caused again to circulate through them, and to circulate normally, its dissolving power owing to the salts being recuperated. In this work the lithium plays only an imaginary rôle.

This explanation in reality applies at the same time to the other two classes of stone consisting of phosphates and oxalates and which may be traced back to the same causes. That is to say the phosphate of potash contained in meat, wine, and beer produces by means of double decomposition phosphate of lime and magnesia with the albumen which is based on iron, soda, lime, and magnesia. In this, however, there is not the slightest danger to our health. A similar decomposition takes place in all cases, even in men who are vegetarians and breathe pure air. The only difference is that vegetarians do not drink spirits which withdraw oxygen from the blood. As soon as we respire sufficient oxygen our "perpetuum mobile" produces sufficient formic acid by means of the spleen and this mixes itself with the blood (see p. 67). This formic acid which is found in the blood and urine of all healthy subjects, dissolves phosphate and oxalate of lime. It is in the interest of the progress of physiology that I am compelled to lay particular stress on this chemical fact, since no mention is made of it in the text books of physiology, pathology and therapeutics. Even Schüssler's "Abgekürzte Therapie" does not make the slightest allusion to this important and necessary constituent of healthy blood. For the present question it is of importance to remember, that the urine of patients suffering from stone does not show the acid reaction of formic acid, but on the contrary shows alkaline reaction, turning red litmus paper blue, instead of turning blue litmus paper red. In this way the precipitation of the phosphates and oxalates and their combination with mucus and gelatine to form "calculi" is rendered comprehensible for the chemist.

It is also of importance from the chemical point of view to bear in mind that oxalate of lime forms with chlorides and sulphates of lime and magnesia double salts, easily soluble and it is just the *sulphates* that are lacking in meat, beer and wine.

Such being the state of the case I can hardly avoid feeling sad, when considering the amount of confusion that prevails in the therapeutics of this question. When patients who have had their urine examined have found that the calculus consists of oxalate of lime, they have generally followed with fear and trembling the orders of their physicians to abstain from nutriment containing lime so as to prevent the growth of the calculi. This error alone would have been sufficient to induce me to write the passage relating to the functions of the spleen and the manner in which it produces formic acid. It must be borne in mind that the albumen of the blood cannot exist without lime. Every moment that we breathe we destroy lime-containing blood corpuscles, whose ashes can only be excreted by the kidneys and bladder. It is consequently physiologically essential that the urine should contain lime. The loss of the blood corpuscles must be made good and hence nutriment containing lime is indispensible. If the patient is not provided therewith, the blood will take the lime which is absolutely indispensable, from the bones. What good is it then to the patient to carefully abstain from nutriment containing lime, so long as he has bones in his body which form for a long time an inexhaustible source of lime! I will give an interesting example:—

Mr. P., a music teacher of Glasgow who in addition to calculus suffers also from pianist's cramp and rheumatism, addressed a letter which came into my possession, to his medical adviser who had been treating his cramp according to my method with iron, sulphur, lime, and physiological salt:—

"About 8 months since I had to endure the passage of a stone, which was one of oxalate of lime and for that reason I was strictly enjoined to avoid any medicine or chemical foods containing the phosphate of lime. A certain syrup of Hypophosphide, known as Gibson's, was ordered on account of its not containing phosphate of lime, though it is strong in iron. You will, I am sure, easily understand my anxiety to keep clear of anything even remotely tending to another agonizing experience. Would it be possible to obtain the medicine without the Lime? Or would there be any substantial difference between Hensel's medicine and the usual phosphate of lime in regard to the possibility of its forming oxalate of lime? — And for some reason I have to avoid sweets. If you should still think the Hensel medicine perfectly safe for my case, or should be able to obtain for me a suitable modification, I shall be very glad indeed, for the treatment seems to me very rational and promising."

This communication insists, that the patient must avoid the use of sweet things. The explanation of this lies in the fact, that the use of sugar when there is an insufficient amount of oxygen in the blood, leads to the formation of lactic acid, which is a product of the decomposition of sugar, and lactic acid acts in a paralyzing manner on the nervous functions. Blood in its normal constitution must, however, have an alkaline reaction in order to make the oxygen breathed effective. The albumen of our blood is essentially a soda-albumen, and in order to give this character to the blood, the physiological salt-water should be used, as this has an alkaline reaction owing to its double-basic phosphate of soda. Its salutary action is supported by the co-operation of combinations of sulphur, lime and iron, which augment the number of the disks of blood formed, and this hemoglobin is the vehicle of the oxygen breathed. In this way two effects are produced, first, the acid reaction in cases of rheumatism, of qualitative poverty of blood, and other pathological states is brought back to its proper alkaline state, and secondly, by a more abundant operation of the oxygen, the oxidation of the rheumatic secretions such as uric acid into carbonic acid, water and nitrogen is gradually effected, this is accompanied with the conversion of the urates difficult of solution, through combination with the physiological salts into double salts easy of solution, which then are discharged unnoticed with the secretion of urine.

As for the rest, there can be no doubt that when *all* the blood vessels are filled with nerve-vivifying oxygenated blood, which also contains the due amount of salts, this must be of advantage to every region of the body. Not only liver, spleen, stomach, pancreas, and intestines, but also the kidneys, the lungs and the skin all work harmoniously together supporting, and complementing each other.

The most important point is the chemical combination of the respired oxygen. When no vigorous stream of oxygenated blood is supplied to the spleen, its function of producing formic acid is inadequately discharged. In this connection it is comprehensible (to make a general application of this particular case), that musicians who for many years have been condemned to breathe the heated impure air full of carbonic acid and poor in oxygen of overfilled concert rooms, are liable to suffer from calculi both of oxalate of lime and uric acid; to this is to be added, that as musicians are always thirsty, owing to the hot air that they breathe, if they quench their thirst with ale, porter or beer, which contain but few salts, their evaporation is increased, and their thirst becomes more violent. Instead of ale, beer and porter they ought to drink mineral water, alternated with tonic lemonade.

I recommended the adviser of Mr. P. all the more to keep to the treatment recommended in this work under the heading of "Asthma", as the complication of calculus, with cramps and rheumatism, which compelled the patient to confine himself to the role of giving lessons on the piano, clearly shows what was in reality the matter, namely the absence in the blood of the *mineral salts*, which give electricity and act as solvents, and in addition of *iron, lime,* and *sulphur* necessary for the new formation of blood corpuscles absorbing oxygen. It must of course be understood that this includes abstinence from alcoholic beverages, as the alcohol would draw to itself the oxygen of the blood which is essential to the proper action of the nerves.

Small-Pox (Pock). We have already seen in the pathological section that in this disease there is a decomposition of the lymph. I believe an inadequate concentration of the lymph to be the cause of this decomposition of the lymphatic fluid. Just as thin beer ferments more easily than one more strongly spiced, so also diluted lymph is more easily decomposed than lymph of a higher specific gravity. A considerable contents of salts and earths is essential to enable the lymph to attain a higher specific gravity and a greater concentration. Salts act as protectives against chemical decomposition; lime, magnesia, manganese, protoxide of iron have a similar effect. Now when the lymph tends to decompose, an insufficient proportion of mineral substances is doubtless the principal cause. A further cause must be sought in the temperature of the air respired, together with its deficiency in oxygen. Now inasmuch as the lymph is extracted from the nutriment consumed, it follows that *inadequate nutriment* and *impure air* must without doubt be regarded as the causes of small-pox.

The patients affected by small-pox whom I have myself seen, were all pale, anaemic and consequently badly nourished. And when we remember that in chlorosis an undue proportion of white corpuscles to red corpuscles occurs, it follows that small-pox critically considered is nothing but modified chlorosis. In the case of chlorosis too the occurence of eruptions and lymph pustules is to a certain degree an attendant symptom. The more this coincidence is considered the more we are inclined to consider small-pox as merely a variety of chlorosis, more especially as the two affections may be prevented and cured by one and the same treatment, namely lime, iron, and "blood-salt".

Children may be effectively protected against small-pox by giving them salted milk, which is also efficacious in chlorosis. The children should be given in each bottle of boiled milk, i. e. in every half pint as much table salt as will lie on the point of a knife. Cow's milk contains per quart about $1/4$ oz. of ash substances, while normal blood contains between $3/10$ and $1/3$ of an ounce. It is therefore advisable to make up for this difference by the addition of salt.

Milk is the most rich in lime of all forms of nutriment, containing in its 7 thousandth of ashes more than the half of lime. Lime protects the albumen from chemical decomposition. When on the other hand the milk for sucklings is diluted with twice its volume of water, the advantages of the lime and iron already pointed out disappear—these bodies being contained in milk to the same proportionate amounts as in normal blood.

Where the nursing mother is of weak build and does not possess enough lime, iron, sulphur, or "blood-salts" for her own requirements, her milk also is necessarily of a watery consistence, and the want of appetite and the appearance of small-pox in her suckling is only too easily explained. I know no better comparison than that of a good strong beef-tea and a miserable weak one. We consume the former with pleasure, we leave the latter untouched.

In the case of adults, tonic lemonade acts both healingly and preventively against small-pox, even when taken alone. I have known of cases in which small-pox had broken out in a family in which the mother had gone through all the preliminary symptoms (shivering, indigestion, vomiting, with pains in the stomach, and the characteristic achings in the back and loins) and still was kept from the actual outbreak of the small-pox pustules by the use of tonic lemonade. Concerning this lemonade see the next article. It is true that a nourishing diet was at the same time used, as soon as the lemonade had restored the appetite. In the case of weakly children half a tea-spoonful of "Hensel's tonic" in a half pint of sugar-water given by means of the bottle will suffice to restore them to a healthy condition of nutrition and digestion.

As far as concerns the prevention of small-pox by the inoculation of small-pox lymph, which strongly reminds us of the days of witchcraft, I must say that I decidedly share the opinion of Herr von C. That is to say, I feel sure, that if a Prosecuting Attorney were to enter the sick chamber with every disciple of Koch, Pasteur, and Jenner, and were to draw up a report in reference to the inoculation and its results, the friends of inoculation would be dissipated like mists by the sun. It is only their freedom from responsibility to common sense and the criminal law, that causes the inoculators to imagine that they are in the right. They ought to be brought back from this mistake by an accompanying Justice of the Peace.

Chlorosis, Anaemia and Irregular Menses. The frequency of these affections in youth where the organism is endeavoring to attain a full developement, is in my experience due to the supply of blood being insufficient to provide for the growth of all the different parts of the body. Especially is the blood deprived of considerable quantities of lime for the growth of the bony structure, and this is converted into bone by combining with the ammonium phosphate of the nerve-substance. This gives rise to two evils: on the one side nerve-substance is used up, and on the other the number of the red corpuscles of the blood is diminished, and these cannot be formed anew as they ought to be in order to make up the loss due to respiration. By this also the protoxide of iron chemically combined with urea, passes from the circulation in the form of uro-hematin in the urine. Just the same thing takes place with the particles of sulphur which take part in the formation of the red corpuscles; they are oxidized and pass off as sulphates in the urine. As sulphur, lime and iron are in this way decreased and the number of the red corpuscles of the blood, which absorb oxygen is correspondingly diminished, the nervous functions must relax. We then find timidity, and want of appetite accompanied by indegestion and constipation. Not only the brain but every part of the body suffers from want of oxidized blood. The tendency to shiver is a consequence of the scarcity of the warming oxygenated blood. The patients feel cold even in the eyes. The resulting want of blood pressure owing to inadequate supply of oxygen causes stoppages producing headache, giddiness, and fainting fits. Cramps, pains in the limbs and rheumatic symptoms also arise from the common cause of imperfect circulation thus produced. Palpitation (owing to paralysis of the controlling sympathetic nerve through want of oxygen), sleepiness and constant weariness (because the blood is overcharged with carbonic acid) complete the catalogue of symptoms due to the one cause of a too small number of the red corpuscles of the blood which absorb oxygen. That in addition the sexual region should suffer either from the general deficiency of vivifying blood or because the blood is pressed by the presence of carbonic acid out of the relaxed tissues in considerable quantities, will be easily understood.

A quantity of the electrifying (exciting) salts leaves the organism contemporaneously with the iron and sulphur that pass off in the urine. This explains why anaemic girls have what is called a "morbid" desire to eat charcoal and chalk. The charcoal contains the ashy constituents that give firmness to the wood as well as to the animal organism, and the chalk is carbonate of lime which is required for the formation of new red corpuscles. It is by no means an indigestible substance. On dissolving in the gastric juice it forms organic compounds which form in conjunction with sulphur, iron and gelatine a basis for fresh blood albumen.

Many cases of chlorosis are easily and quickly cured by some preparations of iron, but in some cases iron preparations produce no effect. The explanation is that without the cooperation of sulphur, and lime, red corpuscles cannot be produced and that without contemporaneous supply of the deficient salts to the blood the digestive juices remain without strength, and assimilation ceases. All these factors must act together to produce success. By adding, therefore, flowers of sulphur and silica even the "fluor albus" which is frequently a concomitant of chlorosis is radically cured.

Of iron preparations the one made according to my receipt and known as *"Hensel's Tonicum"* has proved itself of first class efficacy and has won a great number of friends amongst doctors in all parts of the world. It contains lime and iron protoxide combined with formic and acetic acids there being in every 100 parts 1 part of iron and 1 of lime.*) Its formic and acetic acids enable it to correct the defective action of the spleen and so to break through the "circulus vitiosus" due to imperfect circulation, defective electricity of the nerves, defective production of intestinal juices and impeded new formation of blood and nerve material. So also the paralyzing effect on the nerves of the

*) For North America and Canada the firm "Boericke & Tafel in Philadelphia" is familiar with its preparation, so that a reliable preparation can be furnished by them.

ammonia produced by the congested blood in the capillaries is neutralized by the acids. Its contents in iron and lime enable the white corpuscles which are especially formed in the spleen to be converted into red corpuscles, and this enables the blood to regain with certainty its capacity for chemically combining with the oxygen and so becoming of use for the vital processes of the nerves. At the same time the iron protoxide combining as it does with cyanides prevents further chemical decomposition of blood albumen which manifests itself by the separation of water (hydropsy) and by cyanosis.

Owing to these characteristics "Hensel's tonic" effects a great deal though not everything. It is especially deficient in phosphates, chlorides, and sulphates and cannot consequently replace all the salts eliminated with the urine. In most cases this need not be considered, since the enriching of the blood with oxygen caused by Hensel's tonic, stimulates the activity of the nervous system of the viscera, so that the assimilation becomes normal, and nutrition receives a new impetus and all the necessary materials are obtained from the food. Hence even after a short employment of "Hensel's tonic" the expression is very common "I am as hungry as a wolf. I can eat everything, everything tastes well." It is not, therefore, astonishing that a complete change is visible in 4 weeks treatment with "Hensel's tonic" even in patients that have suffered from chlorosis for months. Girls who appeared at first shy, tearful and as pale as whitewash, often develope even in a month in such a way that they might stand as models for a statue of Germania. They are self-possessed, their forms become rounded, their eye clear, and their steps firm. "Do you know me?" asked a lady who visited me in my office hour. "I have not yet had the honor."—"My name is M. D. you recommended to me the "tonic" a month ago."—"Oh yes! now I remember your face, but you are no longer the same."

"Hensel's tonic" restores to the female body in a relatively short time the cushion of fat which is peculiar to healthy women. This is partly produced by the sugar consumed in the tonic lemonade,*) and by the formic acid**) which is a chemical basis for new fat, and which enables the fat consumed in the food to be better utilized through the production of normal alkaline bile due to the increased supply of oxygenated blood to the liver, thus emulsifying the fats more completely, which are then sucked up by the delicate terminations of the

---

*) Tonic lemonade is made of 4 heaped up tea-spoonfuls of powdered sugar and 1 of Hensel's tonic in half a pint (a tumbles full) of water.

**) A certain physician supposed, that the formic acid used in Hensel's Tonicum was made from crushed ants; but it is now made from oxalic acid and this from sugar. Formic acid $CHHOO$, arises from the oxidation of carbureted hydrogen CHH and is found in many substances, e. g. in pine-needles; it received its name because chemists first discovered it in ants. If it had received its name at this day it would perhaps be called primary acid, because it occupies the primary position among fat acids.

lymphatics in the alimentary canal. The lymphatic system fills itself twice in 24 hours with new juice. If even one night of strengthening sleep can restore fulness to the cheeks, how much more 30 whole days!—

Nevertheless, as already pointed out, there are some cases of chlorosis that cannot be cured by the sole use of "Hensel's tonic", owing to the fact that the patient's blood has in the lapse of time become too watery i. e. poor in salts. Such patients can then derive no gain from milk, as they are incapable of digesting either the caseïn or the butter fats it contains. It is consequently advisable to ask such patients if milk and eggs agree with them. If they do not agree with them, these natural forms of food must be replaced by beef-tea in which the yolk of an egg has been beaten up; this by means of its contents of phosphates of potash and soda from the meat, and the ordinary salt which has been added improves the constitution of the salty serum. This method of introducing into the circulation fresh serum through the lymphatics, appears so natural, that one can only reserve a melancholy smile for the antiquated and often fatal method of attaining the same end by the transfusion of blood from a healthy man into the artery of a weakly invalid. Even the previous removal of the fibrin from the transfused blood hardly makes the method more rational. If it was in some cases successful, this should be ascribed to the electrifying salty contents of the serum, to which the worshippers of bacteria even ascribe the power of destroying bacilli, and indeed not incorrectly; for the so-called bacilli disappear when, owing to a sufficient proportion of salts and earths, normal albumen is formed and disintegration ceases. Since this is done best in the way instituted by God, by allowing the transfusion to take place through absorption by the lymphatics, I affirm that the previous removal of fibrin would not make the violent injection of blood into the opened arteries any more rational. Alas! What innumerable absurdities will ever be perpetrated, so long as people continue to ascribe mysterious composition and properties to the albumen of the blood through sheer lack of chemical knowledge whereas the blood-albumen consists merely of a combination of partially oxidized and partially unoxidized sugar with ammonia, alkalies and earths whence it assumes an electro-negative relation to the serum, with its chlorides, sulphates and phosphates of alkalies and earths heightened in its electric force by its formic acid and which owing to its formic acid acts as a solvent.

The number of cases of chlorosis and anaemia which have been successfully treated with physiological salt-water either alone or mixed with equal amounts of warm milk, alternating with tonic lemonade and in combination with sulphur and lime are so numerous and "Hensel's tonic" has become so generally known, that I need give no examples. With respect to the assistance of silicic acid in effecting a radical cure of *fluor albus*, it may, however, be of interest to point out that there are constitutions which are so thoroughly devoid of silica that even small quantities of it give rise to powerful reactions in a body entirely unaccustomed to it. The following correspondence may be found instructive with reference to a mother and daughter both of whom suffered from imperfect constitution of the blood.

M. July 8, 1890.

My daughter had influenza some 4 weeks ago. After this she got a tetter on the hand. Small red knots formed under the epidermis, which in a few days filled with a watery fluid after which they became covered with a scab covering nearly half the hand. Similar knots soon appeared on adjacent parts of the skin hitherto spared. Eight days ago also the lower arm and the face were attacked while shivering fits came on. Quinine pills and then arsenic were prescribed by the physician who regarded the tetter as sequela of the influenza. Later on mercurial ointment was externally applied on a part of the lower arm which was healing, but with very doubtful results. Please advise.   E. S.

(I recommended the patient to wait 10 days longer and to await further developments, and then to advise me again.)

M. July 20, 90.

I must proceed to inform you that my daughter's complaint has assumed a different character. The cause according to our physician is an attack of diphtheritis which she had last week. The face in time healed without suppuration, but small sores formed on the back of the hands with the appearance of fever (103.6⁰). These also extended to the palms and fingers and in the end they coalesced into one sore. The doctor ordered external application of sublimate, after the fever abated; after an application of 8 hours, which finally caused great pain, some amelioration seemed to set in and the formation of pus ceased in some places. We must wait and see if a complete cure will result.

M. July 21, 90.

I must now inform you that the fingers have again begun to fester this morning in spite of their being treated with the sublimate. What do you advise now?

(I recommended a glass of tonic lemonade in the course of the morning and another before bed-time; also milk with 1 tea-spoonful of salt per quart of boiled milk. For the suppurating hands I recommended washing i. e. cleansing with a decoction of camomile and wax-salve, the milder the better. The tannic acid in the camomile coagulates and disinfects the disintegrating albumen of the blood.)

M. July 23, 90.

— — After putting on the sublimate, great pain was experienced and all the sores showed a constant flux of blood, water, pus and a milky fluid. I commenced yesterday at once the washing with decoction of camomile, and also the wax-salve recommended by you and an amelioration is already visible and the large sores look somewhat better, but the festering continues, and little pustules filled with matter are appearing on parts of the lower arm hitherto spared.

M. Aug. 3, 90.

Your treatment of my daughter has been a success. The large sores have closed and there is a very strong scabbing off. Small pustules however still form on the new skin.

I myself have had "fluor albus" for two years since my menses ceased and am often very weak. May I ask for your advice?—

124

(In this case also I recommended 1 to 2 glasses of tonic lemonade per diem and in addition $\frac{1}{2}$ a pint of physiological salt water, reserving flowers of sulphur and calcium magnesium phosphate in case the above should not suffice.)

M. Aug. 24, 90.

On July 22. you ordered my daughter daily 2 glasses of tonic lemonade 1 pint of salted milk with flowers of sulphur in case that a cure should not be effected within 3 or four weeks. This produced very great improvement, the sores closed, the skin on the arms and hands became paler, the dead skin began to peel off to our great joy,—when—5 or 6 days ago a relapse set in. The skin again became red, and small pustules formed upon it which itch strongly and contain a clear liquid. I cannot explain it, as our manner of living has in no way been changed. Do you think we ought to resort to the flowers of sulphur and the other medicaments?

As to myself, my general condition is a little better, but the local affection has not been altered. My former physician had ordered injections with water and $1\frac{1}{2}$ table spoonsful of wood vinegar. They diminished the discharge, shall I recommence them? The physician suggested a possibility of cancer, but I shall not submit to an operation.—I have suffered for two years but have not had pains. I have no swelling. I have not submitted to a local examination.

(I prescribed a daily dose of flowers of sulphur for both mother and daughter in addition to the tonic lemonade and also $\frac{1}{4}$ of a tea spoonful of calcium magnesium phosphate in milk salted and sugared, twice a day. There was no objection to the local application of the wood vinegar.)

M. Sep. 16, 90.

I have accurately followed all your directions for my affliction (fluor albus), and my general condition is better than for the last two years, but the local affection still persists, I cannot altogether avoid in my household a certain amount of work and mental excitement which may be prejudicial to my recovery.

With my daughter the sulphur has put a stop to the appearance of new pustules, though the skin on arms and hands is still very sensitive. The skin gives the impression of being liable to break and probably several weeks will elapse before the skin will regain its former softness and its natural color.

M. Oct. 5, 90.

Your prescriptions besides improving my general health have also though slowly diminished the local affection.

My daughter who still follows your prescription is delighted with the continual improvement which has taken place, especially as many here have been similarly afflicted without finding a cure in arsenic pills and quicksilver.

M. Nov. 22, 90.

All your prescriptions have benefitted me and my daughter, but we are still subject ever now and then to relapses. My daughter who was prevented from singing by hoarseness was examined with the throat-mirror and was found to have inflammation of the false vocal cords. She was recommended to inhale morphine and glycerine and combats the inflammation besides with tannin. After inhaling, my daughter found that her hands began again to itch. If she leaves off the morphine she again becomes hoarse. Etc.

M. Feb. 4, 1891.

I should like to repeat from time to time and to continue the treatment you recommended for keeping the blood in good condition as it is years since I have felt so well as since using your remedies.

I have suffered for years from "pins and needles" in the hands, which increased when knitting or sewing &c. I had also great pains in the wrists, often so violent that I did not like to move them. All this has disappeared almost completely and the "fluor albus" is also decidedly better. I dont know if there is any connection between these complaints but I am delighted at the cure.

(I recommended as the fluor albus still continued, to try ¼ of a tea-spoonful of amorphous silica stirred into ½ a cup of boiled and salted milk.)

M. Feb. 12, 91.

In thanking you for your advice I must inform you that the silica which I took as prescribed (¼ teaspoonfull in ½ cup of milk) disagreed with me so decidedly that I had to give it up. Only now after 5 days am I at all right again and shall resume the old cure with sulphur and phosphates &c.

A quarter of an hour after taking the silica I had a feeling of fulness followed by vomiting, purging headache and entire loss of appetite, which has only returned to-day. As my general condition was under the circumstances so well before, I can only ascribe the change to the silica. This strange effect may be of interest to you.

(I replied that such a crisis produced by silica was indeed unusual, and showed that the whole system, the stomach as well as the intestines, the head and the abdomen must be almost devoid of silica, which considering a 50 years' diet of white branless bread is easily enough explained. Nevertheless the silica is a necessary constituent of the system existing even in the brainfat and the yolk of eggs and is in no way a poison. The patient therefore should make another attempt, taking at first only a small pinch of silica daily and so gradually accustoming the system altogether unaccustomed to it, like you would a little child. Without silica no firm bodily tissue can be produced.)

M. April 23, 1891.

The continued employment of silica in accordance with your advice has completely removed my complaint (fluor albus). I take a quarter of a teaspoonful daily for some time already and wish to continue this cure together with sulphur as I have not felt so well for years.

My daughter who has also been completely cured by your treatment looking to the improvement of the constitution of the blood, intends to keep to the tonic lemonade as a health preserving beverage."

Experience shows that the tonic lemonade as a beverage, (a glass every other day) is a preventive against many diseases, e. g. small pox, diphtheria, scarlet fever, measles, scrofula &c. always assuming that the nutriments contain enough sulphur, lime, and silica; otherwise these substances also must be supplied to the blood.

I will conclude this section by a short notice selected from among many hundreds reported to me:—An anaemic female patients was sent by her doctor to Italy to recover in a milder climate. She, however, only got as far as Kufstein where she fainted and had to be taken from the train to the house of a switchman. The railway doctor was summoned and as he had had experience with "Hensel's tonic" and always carried a small bottle of it with him, he gave the almost lifeless patient a glass of the tonic lemonade. Her eyes thereupon became bright and the whole body was revived. After remaining a few weeks in the house of this doctor, she was so completely restored to health by "Hensel's tonic" and strengthening diet, that she declared—"I was going to Italy to get well. As I am now in good health I need not any longer go there. Please give me a reserve of "Hensel's tonic" and I will go home to my own."

Cholera morbus.—Cholera.—Dysentery. These diseases agree in being characterized by a paralyzed state of the intestinal nerve which is connected with the nerves of the skin and largely dependant on their function. This connection is explained by the developement of the embryo,—the terminations of the nerves passing to the outer skin and to the inner epithelium, forming a connected whole, owing to their common origin from the outer layer of the embryonic germ mass with its three layers. Just as the one half of the *outside* covering of a flat closed bag may by sowing together the opposite sides become the *inner surface* so the external layer of the blasto-derm of the embryo becomes the serous inner coat by the growing together of the edges, a visible sign of which is afforded by the "linea alba" of the peritoneum. We also learn from embryology that the vascular system is developed from the "lamina media" of the embryonic germ mass, whence it is called "lamina vascularis" while the innermost "lamina interna" giving rise to the glands and other organs lined with epithelium is termed "lamina mucosa" or "intestino-glandularis." This inner region enclosing the intestines is anatomically closely connected with the sensitive external layer because the inner covering (epithelium) produced from the latter, extends over the mucous membrane with the closest application to the same. It is owing to this anatomic structure that the inner and outer serous membranes must be regarded as the external and internal poles of a connected whole. Hence it comes that the outer skin can be brought to perform its functions properly by internal means e. g. by sudorific substances, as is done in practice; but the paralyzed state of the internal coat may also be removed by external applications.

The knowledge of this connection affords a clue to the treatment of conditions produced by external atmospheric conditions, such as cholera morbus,—cholera, and dysentery.

The occurrence of cholera morbus in young children brought up on the bottle is most prevalent in the hot summer months. Previous perspiration disposes to chill of the outer skin resulting in a partial paralysis of the nerves of the skin which communicates itself to the nerves of the inner mucous membrane. The result is that the glands do not produce the proper digestive secretions. The bile, the pancreatic juice and the intestinal juice become weak, since the paralyzed termin ations of the nerves have lost their power of attracting the oxygenated arterial blood. The interruption of circulation in the capillaries of the walls of the stomach shows itself by vomiting, and we may draw conclusions in regard to the same state in the capillaries of the whole mucous membrane of the intestines from the state of the mucous membrane of the stomach, since the stomach is nothing more than the commencement of the intestinal canal with widened walls. Since wherever blood stagnates in the capillaries nerve action refuses its service, and since on the nerve action depends the secretion of the digestive juices, (the gastric juice, intestinal juice, gall and pancreatic juice) we understand why in the case of cholera morbus the casein separated from milk in the stomach is not digested. We find it unaltered in the faeces that imitate the appearance of hard-boiled white of egg cut up and mixed with the green coloring matter of the bile, biliverdin, showing that the gall bladder also is affected by the convulsive state to which all the branches of the nerves of the viscera are subjected, because the circulation and renewal of the blood is impeded in the neighboring capillary vessels. For if the capillary blood renewed itself properly and brought fresh oxygen to the gall bladder the green coloring matter would be oxidized and the stools be of a golden yellow color.

For curing the cholera morbus of sucklings two methods may be simultaneously adopted. One is to rub the whole body with weak, lukewarm vinegar, i. e. common vinegar and water in equal parts. This operates as an electric stimulus upon the nerves of the skin, and this effect is communicated to the nerves of the mucous membrane of the intestines. The second is to give precipitated carbonate of lime— 8 grains in a spoonful of sugar water as often as the diarrhoea or vomiting return. As a rule two or three doses suffice. The lime gives a new consistence to the lymph and the prevailing state of fermentation is stopped. Earth, calcareous earth gives fresh energy and keeps the little ones above the ground. On the other hand opium which has still its adherents, has simply the effect of intensifying the paralyzed state of the nerves and of bringing the suffering children under the ground with tolerable certainty.

In addition suitable nourishment not as liable to decomposition as milk and capable of giving body to the watery lymph acts with visibly rapid restorative effect. Such a nutriment is the gelatinous broth made from calves feet by boiling them in water with some salt.

The dysentery of adults has this in common with the *cholera morbus* of infants that in consequence of a peculiar electrical condition of the atmosphere it attacks numbers at once. On the other hand the disappearance of dysentery as soon as the atmospheric conditions revert to the normal, shows it to be allied with *cholera.*

A convulsive state is characteristic also of dysentery, the characteristic symptom of which is a constant urging to evacuation while but minute amounts of faeces are passed. Owing to this similarity I have consequently tried for dysentery also, a treatment of rubbing the body with lukewarm vinegar, and the internal administration of carbonate of lime in doses of $1/1$ of a teaspoonful in sugar water, but I also gave at the same time 8 grains of flowers of sulphur twice daily. As sulphur is electro-negative in its character, its quieting effect should be ascribed to its allaying the convulsive state of the electro-positive nerve substance of the mucous membrane.

According to this theory the same treatment, namely external application of vinegar, and internal administration of lime and sulphur with the addition of formate of iron (Hensel's tonic) for the purpose of forming fresh blood corpuscles, is to be urgently recommended in cholera. The employment of sulphur is to be specially recommended because any excess is carried off by the activity of the skin in the form of sweat— a result due to the renewed activity of the terminations of the mucous membrane. It is with such a revivification of the nerves of the skin that according to experience the convalescence in cholera patients commences;—when a warm perspiration breaks out they may be regarded as saved. That salted calves-foot broth should be used in cases of cholera to make up for the loss of lymphatic fluid, is to be considered simply as rational.

Catarrhal Affections. The nature of these affections consists in a throwing off of the fine membrane which serves as a protecting covering to the serous layer which conveys the lymphatic juice. Owing to the fact that the epithelial cells separate themselves from the serous layer, which contains imbedded in it the terminations of nerves, capillaries and lymphatic vessels, the parts exposed namely the nerves, the plasma of the blood and the lymph are subjected to chemic alteration owing to the oxidizing action of the air. The result is seen in products of decomposition sometimes in large, sometimes in small amounts. Like all other products of decomposition and fermentation. the catarrhal secretions also possess an infectious character inasmuch as the decomposition .of which they are the products is transmitted to healthy portions of the tissue, an example in point being furnished by the ordinary cold which as experience shows may, when precautions are neglected, be transmitted from one member of a family to another.

As in the case of an ordinary cold in the head which is not cured at first, the first watery secretions are followed by tougher yellow mucus and accompanied by swelling and redness—a proof of stagnation of the capillary circulation, similar processes take place in all affections of a catarrhal character. Not unfrequently the chemical decomposition to which the stagnant blood is subjected actually produces suppuration, i. e. the destruction of the connective tissue in which the blood-vessels and nerves are imbedded, and in this way owing to the absorptive action of the lymphatics which distribute the septic matter throughout the system, there arises the possibility of the formation of the state designated as "cancer", and which cannot be effectively combatted by local operations, since the lymphatic system pervades the whole body and so conveys the germs of decomposition in all directions. Such being the case, it is plain that Emperor Frederick could not be cured by local scratching or scraping of his cancer which had been developed from a catarrhal condition which owing to the paralyzing action of the ammoniacal nicotine upon the epithelium of the air-passages had gained a headway. Nor could the generous diet of milk and eggs recruit his strength, since the lymphatics of the whole system were infected with a matter bearing the germs of infection.

The only treatment that can effect a cure consists in bringing back the condition of the external and internal serous membrane full of nerves and containing the lymph, to the normal and arresting the decomposition of the substance of the connective tissue by the action of electricity.

Of a value equal to that of electricity is the warmth due to the rapid circulation of the blood, and so we see that at the commencement of a catarrh the introduction of warmth by means of a hot drink will effect a cure. Conversely imperfect circulation which accompanies a defective constitution of the blood leads with women to catarrhal affections which are at first regarded as local complaints, but subsequently as cancer pass into a general decomposition of the lymph. From this point of view the so frequently occuring utero-vaginal-catarrh (fluor albus) is often found to be the first stage toward uterine cancer, cancer of the breast, and also of hepatic and gastric cancer, a fact which when we take into consideration the connection of the organs by means of the branches of the sympathetic nerve and the net-work of the lymphatic system, cannot surprise us. Leaving these facts however, out of consideration, it is of course plain that the vascular system also contributes to the spreading of catarrhal secretions throughout the whole body, for whatever catarrhal decomposing matter is taken up by the lymphatics they pour into the stream of blood between the subclavian and the internal jugular veins. In this way it may be understood, how the cancer is constantly supplied with fresh cartilaginous substance, which is transported throughout the body and deposited at unsuitable places where it gives rise to indurations and ulcerations, which tend to ulcerating decomposition and to hemorrhages.

Whence the cartilaginous substance that produces the ulcers comes, is evident from a consideration of the relations of the lymphatics which inside and outside of the thoracic cavity nestle close to the ribs and take up from these the cartilaginous substance. This passes from the lymphatics into the veins and from the veins into the arteries and thus the "circulus vitiosus" is completed, through the co-operation of the internal mammary artery from which branch off the arteries supplying the bronchia and the diaphragm, and the anterior intercostal arteries, as also the superior epigastric which in the neighborhood of the navel unites with the inferior epigastric, which is itself a branch of the femoral artery. Now when we take into consideration that the inferior epigastric supplies the uterus and the ovaries, that it is connected with the superior epigastric which forms a branch of the internal mammary, giving off the intercostal arteries which in turn dispatch the exterior mammary arteries to the mammary glands, we are compelled to ask whether those surgeons who entertain the hope of curing cancerous ulcerations by excision are really quite clear on the anatomical points involved—particularly the connection between the mammary glands, uterus, and ovaries?

Surgery ought to have been startled long ago at the fact that cancers in the lymphatic glands when removed with the knife give rise to more energetic tumors than before; but it seems to be the case that when a surgeon takes up the knife, he expects it not only to cut for him but to do his thinking as well. Cutting away certain parts of the body in which the cartilaginous material that has been dissolved by the lymph happens to have deposited itself can never stop the process of chemical decomposition that passes by the name of cancer; that can only be properly effected by normalizing the composition of the blood by the action of disinfecting or antiseptic salts—by iron, lime and sulphur which help the decomposing albumen to regain its normal condition and soundness. Blood which is thus normally constituted takes up sufficient oxygen which is of course a disinfectant through its oxidizing action and in this way at the same time the main object is attained:—namely to bring into action nerve force, vital warmth and the movements of the blood. In this way one may in great part guard against and make oneself independent of the harmful influences of the weather which find their expression in colds, grippe, influenza and bronchial, gastric, and intestinal catarrhs, as well as catarrh of the bladder; and now and then as Bright's Disease.

That Bright's Disease also may be considered in this sense as catarrhal in nature in so far as it shows infections of septic material, is shown by the secretion of water in the serous membranes in the form of dropsy. As chilling of the air-pipes gives rise to cold and bronchial catarrh, and cold drinks may occasion gastric and intestinal catarrhs, so a chill in the back in the region of the kidneys may by paralyzing the renal plexus give rise to renal catarrh. Owing to this paralytic state a certain portion of the albumen passes over into the urine which symptom is just as characteristic of renal catarrh as is the presence of albumen in other catarrhal secretions. In all these cases the paralyzing action on the nerves, of the ammonia which is disengaged from the stagnating blood of the capillaries appears to be an important factor in introducing the affection, just as the inhalation of gaseous ammonia instantly produces nasal catarrh. In agreement with the recognition of ammonia as the cause of the mischief, lemon-juice in the form of lemonade is employed in Bright's Disease; as a cure for other catarrhs hot sugar-water with plenty of lemon-juice with rum is customarily used. It is therefore not to be wondered at that that sailors and the coast population from Holland to Memel, in England, Sweden, and America, and also the inhabitants of Russia, and generally all who are exposed to dampness and cold, are accustomed to employ hot grog as a *preventive* against cold, or else alcohol, which being easily oxidized by the respiration of cold air rich in oxygen, assists in keeping up the bodily heat. From this point of view nothing can be advanced against the use of alcohol in bad weather as a prophylactic and curative factor. It is altogether another question when we consider the habit of drinking alcoholic beverages in warm weather or in a heated room. In this case we are not only breathing thin air, deficient in oxygen but we also deprive our blood of the oxygen that is required for vivifying the nervous system by the alcohol contained in the beverages; while in addition owing to the increased secretion of urine we carry off blood salts from the organism. To these two causes—diminished contents of oxygen and of salts in the blood—many cases of gastric and intestinal catarrh may be traced and these in turn are traceable to excessive indulgence in wine and beer. We must again state that bears and seals not only do not feel cold, but that they do not suffer from colds, catarrhs, influenza, or coughs. This is owing to their constant respiration of cool air and drinking of salt water with the fish that they consume. Salts act as electrical excitants and also produce warmth. "Salt makes fat", says the proverb. In this connection I refer to the capacity for assimilating fats possessed by the bile, which is generated by chemical changes from the gelatine and salts of the blood, when the latter is adequately supplied with oxygen by the hepatic artery. This depends upon the sufficiency of red corpuscles which absorb oxygen and these presuppose an albuminous substance built up on a basis of sulphur, lime and iron. In this manner we possess by taking foods containing sulphur, lime and iron in combination with salt, the means of storing up in the system a certain quantity of fat; but in this substance we possess a source of electricity and of vital warmth, whenever it is consumed by respiration; for every substance when it combines with oxygen becomes a source of electricity and is consequently a sort of guarrantee of health and of immunity from the prejudicial effects of unfavorable atmospheric conditions. So much the worse is it that the wine-drinkers remove from their body the foundation of the strength of their blood and nerves, (sulphur, lime, iron and blood salts) and with the mistaken idea that they increase the enjoyment of life, they are only too often hastening towards an early death.

In all classes of catarrh we are to consider as remedies: physiological salt-water, together with preparations of sulphur, lime and iron.

Diabetes. I have already given the necessary information as to this complaint in my work "Das Leben" page 433. It is in all cases a question of insufficient bodily electricity, while the blood in almost all cases is deficient in oxygen. Everything which acts as an electric excitant, such as rapid respiration, rubbing with vinegar, and salt water containing lime and iron, all these tend to prevent the fermenting decomposition of the blood albumen of the liver into sugar and urea; but also the electricity which is peculiar to the soil on which we walk, and which is transferred to our body is possessed of wonderful healing properties. In this respect Carlsbad has a world-wide reputation. How close to the hot spring itself there rises a wildly romantic group of rocks whose many-clefted huge masses bear impressive witness to the volcanic powers within the earth, which have raised them up! Warmth and electricity are of equal value. Where the hot spring bursts forth from the interior, the earth is in a state of electrical tension and this electricity transfers itself not only from without to those who walk above the spot but also from within while they drink the electrizing salt water containing lime and iron. The electricity keeps the bodily substance together and protects it against premature decomposition. Hence Carlsbad regulates all the bodily functions which are subservient to the processes of new construction. All the glandular organs (liver, kidneys, spleen, stomach, intestines and uterus) are anew enabled to perform their functions normally. It is true, that the effect diminishes as soon as the patient leaves that electrical soil and hence Carlsbad witnesses every year a return of the same patients. Is it not a natural conclusion to come to, that we ought to strive to transfer the constant source of electricity into our own blood?

I treat diabetes with nerve-salt and hematite, (sesqui-oxide of iron). The nerve salt produces in the liver active nerve-fat (lecitin) and at the same time it produces electrical tension in all the branches of the abdominal nervous system. In this way these nerves exercise an attraction upon the blood, provided that this blood contains sufficient attractable material viz. hemoglobin. The latter is supplied by the hematite.

I prescribe 8 grains of hematite i. e. sesquioxide of iron, in the form of a lozenge or in a gelatine capsule immediately before dinner; and a tea-spoonful of nerve-salt dissolved in a wine glass of water at 10 in the morning and 5 in the afternoon. A diabetic patient who was thus cured writes as follows:—

Berlin, July 20, 1891.

Some years ago I had the pleasure of hearing your instructive lecture in the Mechanic's Union before the Halle-gate and of reading its contents in the pamphlet which subsequently appeared. The salt-water therein prescribed I took for some time with the best results, but after a while I became indolent and discontinued it. I now know that this was a great mistake. Feeling very weak, I remembered the diabetes spoken of in your pamphlet. I asked myself whether I had not perhaps got it, and on having an analysis made at the Alexandrinen-Pharmacy, 3 per cent of sugar was discovered, the sp. gr. being 1.025. I then ordered the nerve-salt and hematite recommended in the pamphlet; and took it regularly as advised. On the 29th. of June a second analysis gave

Physical appearance: light yellow and clear
Reaction: acid
Sp. Gr.: 1.016
Albumen: none
Sugar: ⎫
Biliverdin: ⎬ none.

I am now not only free from diabetes but am quite a different man, having recovered my former strength, cheerfulness, and power of tranquil thinking. Before this I became tired after an hour's walk, but have just now been walking 10 hours a day in the Saxon Switzerland without being in the least tired.

Dear Sir. I am a man of 40 who have still to support my family, and I cannot sufficiently thank God, that there are such effective remedies for this dangerous disease and I hope that they will soon become widely known.

Etc.   E. A.

Diphtheria. This complaint among children first shows itself plainly by a catarrh of the mucous membrane of the back part of the mouth and of the nasal cavity; it is generally brought on by inhaling cold and raw air (in winds from the north-east), but like all catarrhs which are characterized by infectious decomposition products it may also be brought on by infection. It is a peculiar fact that children who received a large amount of meat and appear to the eye to be robust and strong, more readily fall victims to this complaint than the weaker looking children of poorer people. The cause is doubtless the same as that which causes adults who give preference to a meat diet to be more liable to catarrhs than vegetarians. Lean meat which is in summer exposed in the markets for sale commences very rapidly to undergo chemical decomposition: and the products of decomposition possess a poisonous character as the meat contains nearly a quarter of its weight of cyanogen compounds; the remaining three quarters consist of the constituents of water. How different in this respect are milk, cheese, eggs, cereals and potatoes, the chemical basis of which is sugar, starch, fat and albumen, and even this albumen though it contains nitrogen is not based on cyanogen compounds but upon sugar and ammonia. The cyanogen compounds of meat are produced from albumen by the chemical separation of water. Hence the vegetarian diet in opposition to the carnivorous is to be recommended for children as a protection against diphtheria.

I have frequently in the course of these remarks made the statement that all stagnating blood gives off ammonia. The proof of this may be obtained by putting blood fresh from the artery into a saucer and covering it with a sheet of blotting-paper saturated with nitrate of mercury. It may then be observed that the paper becomes blackened by the sub-oxide of mercury which the rising ammonia liberates from its combination with nitric acid.

The case is the same in regard to the blood which stagnates in the capillaries. As soon as the *mechanical* movement of the blood ceases, a *chemical* movement in which ammonia is liberated takes its place, or if ammonia is not set free then compounds of ammonia with carbohydrates, the so-called amines are formed—a fact which has been illustrated by the grouping of atoms when treating of the decomposition of creatin (see page 67). That a chemical decomposition of this sort which even under ordinary circumstances in the case of dead meat is rapidly infective, must go on even more rapidly where the circulating blood carries the products of infection along with it, can excite no surprise. Hence the greater number of cases of diphtheria in which a doctor is called, terminates fatally, for the simple reason that he has been called in too late. The poisonous putrefaction of the blood has by that time usually made so much progress that the antiseptic remedies which might have been efficacious at first, come too late. This is the ground of my statement, that the treatment of diphtheria must be relegated to the family and not to the doctor. The course of the disease is much too rapid to permit of the doctor being called in, especially in cases occurring in the country where men have to send for miles for a physician, and where the people, therefore, of necessity must depend on themselves.

The numerous medical theses on diphtheria and its treatment are of no practical value, as they shed no light on the real nature of the disease; nor can they, for the simple reason that the authors themselves are not clear about it.

What alone can guide us as to the way in which diphtheria should be treated are the products of putrefaction formed in the ulcerations of the throat. Putrefaction always takes place after the cessation of electricity; as it is this latter which holds the flesh compounds together inodorous during life. From this it follows that it is by supplying new electric tension, that we must proceed to stop the process of putrefaction. Everything serving to set electricity at work tends to effect the cure of diphtheria. Principal among these is the source of all bodily electricity—unimpeded circulation. Packing in wet cloths is thus of advantage as it works as an electrical excitant upon the terminations of the nerves of the skin, and draws the blood from the regions where it stagnates, thus bringing the stagnating blood into motion. Of still greater efficacy than mere water is vinegar. If the whole body be rubbed down with vinegar on the first appearance of diphtheria (difficulty of swallowing, peevishness, loss of appetite), the patients are at once brought out of danger. Vinegar acts as an electrical excitant like all acids. Meat is protected against putrefaction by rubbing it with vinegar, just as cherries are kept from fermentation by pouring vinegar over them; this is so old an experience, that we cannot wonder at the protecting and healing effects of rubbing with vinegar in cases of diphtheria. This remedy is so simple and at the same time so efficacious that nearly every thing else becomes superfluous. But it may be admitted that there are also other factors which have the effect of stopping the process of putrefaction and of restoring the circulation to its normal condition, and are, therefore, effective. To these belong the aperients formerly employed, especially enemas of castor-oil and chamomile-tea, since these liberate the blood collected in the abdominal arteries for purposes of digestion; so also emetics were formerly employed with good effect, as they enable the blood in the walls of the stomach when this organ has been emptied, to take its place in the general circulation. In like manner calomel exercises an electrically exciting effect on the intestines since both chlorine and mercury have the effect of protecting albumen from chemical decomposition. In this connection the calomel of the allopaths in small doses is to be recognized as a pre-eminently disinfective substance.

Even the officinal chlorine-water diluted with four times its volume of water and administered every quarter of an hour in a porcelain spoon (silver cannot be used as silver renders the chlorine ineffective, owing to the formation of chloride of silver) has produced very good results. Hydrogen per-oxide is a remedy of the same class and of similar efficacy and should be administered with a porcelain spoon every quarter of an hour.

Even lime-water painted on the ulcers or administered internally tends to oppose the decomposition of albumen and hence is not unreasonably employed as an antiseptic and a disinfectant.

Acetate of ammonia has a similar antiseptic action. Even small doses of half a tea-spoonful exercise a stimulating action on the circulation as is shown by the breaking out of a warm perspiration.

More remarkable still is the action produced by Hensel's tonic employed as a beverage, a tea-spoonful in half a pint of water, with four piled up tea-spoonfuls of cane-sugar; it acts both antiseptically and owing to its magnetic iron as an electrical excitant. Where this preparation has been introduced into families and the children are given a glass of it a couple of times a week they have invariably been protected against diphtheria. Hensel's tonic has also shown itself able to save life in cases of the greatest danger from diphtheria when the doctor who had been called in has happened to be provided with it; numerous instances of this have been published especially in homoeopathic journals. The preparation cannot be said to be allopathic but rather physiological, and at the same time homoeopathic, for since it contains exactly 1 per cent of iron it corresponds to the second decimal dilution of ferrum and the lemonade made from this dilution 2.5 grammes per 250 grammes of sugar water is also in a proportion of 1 to 100, thus the preparation as a matter of fact agrees with homoeopathic principles. The effects it has produced are already aknowledged in all the quarters of the globe. Finally physiological salt-water also given as a beverage in diphtheria secures a rapid cure.)

Inflammation. At the meeting of the "Gesellschaft deutscher Naturforscher und Aerzte" in Halle (Sept. 1891) Prof. Nothnagel of Vienna delivered a lecture entitled "The Limits of the Art of Healing". He therein defined healing as: bringing back disturbed functions to their normal condition and the stoppage of the pathological processes. Doubtless this is the essence of healing, but this cannot be effected by the physician but only by the blood in circulation. *"Medicus curat, natura sanat"*. The physician has only to take care that nature may be put in a position to exercise her healing power. In regard to *inflammation* Prof. Nothnagel is of opinion that *rest, application of cold,* and *local bleeding* still preserve the efficacy they have been for centuries supposed to possess, wherever their application is practicable: but this method is not applicable where the inflammation is deep-seated as in inflammation of the internal mucous membranes.

It must here be stated, that this position of Prof. Nothnagel is an erroneous one. It is indubitable as shown in my explanation pp. 45 to 48, that as in swoons a species of internal bleeding may take place from the upper part of the body into the capacious venous system of the abdomen, so also in the portal system of veins, in the sinuses, and in the large veins running along the spinal column, the blood which is stagnating and tending to over-heating and inflammation may be deflected to the surface of the body by electrizing the nerves of the skin by rubbing with vinegar. Exactly the same thing may be done when the blood congested in the capillaries of the epithelium tends to inflammation, which expression we shall have to examine, however, more closely. There can be no doubt that Professor Nothnagel is incorrect in supposing that blood cannot be withdrawn from more central parts of the organism. This is most certainly possible, though cupping or bleeding and allowing the blood withdrawn to run out into a basin is not necessary. All that is necessary is to direct the blood to other parts of the body where it can join the general circulation and representing vital power, may do good. Have we studied anatomy merely to forget the way in which the whole vascular system coheres as a whole, and that under normal conditions the blood passes from the aorta to the capillaries and thence through the venous system back again to the heart? And with such a simple condition of affairs ought it to be an impossibility to remove the stagnating blood from the epithelium so as to enable fresh, healthy blood to take its place, and to act in a curative manner? It is just this fact that has enabled the so-called "Nature Doctors" to work seeming wonders with their hydropathy, since they employ the whole surface of the body as a temporary reservoir for the blood, and so have gradually and imperceptibly taken the reins and the dominion away from the physicians of the schools who view the "cells" and "tissue" as diseased and to be cured.

We may agree with Prof. Nothnagel in his doubts as to whether a process like that of Robert Koch would effect the desired end of reconstructing the diseased portions of the organism by its own action, and we may agree with him when he says that the more complete the physician's knowledge becomes, the more he comes to see that *the physician is the servant and not the master of nature.* Or as already said *"Natura sanat, medicus curat."*

All this goes to show that the error which Prof. Nothnagel has proclaimed more openly and loudly than had ever been done before, that the inflammatory blood of the capillaries cannot be easily drawn away, is much more generally diffused among the physicians of the schools than might have been supposed; this explains the lack of success of the old school and the manifest results obtained in many cases by Kneipp, ignorant of anatomy as he is, by means of his merely instinctive and one-sided mechanical procedure (upper shower-bath, lower shower-bath and thigh-bath).

Not that I would recommend a journey to Woerishofen from which place I have within a short space of time received 4 letters from patients who have *not* been cured; it is merely that I wish to draw attention to the powerful reservoir of electricity which is at our disposal in the surface of the body, provided that we bring the nerves of the skin into action, which is useful in all cases of stagnation and inflammation of the blood. Both acids and salts act as electrical excitants. Salt-water may therefore be employed for rubbing the skin, instead of vinegar, in cases of inflammation; the salt-water should be made of 38 grains (a tea-spoonful) of salt and ½ pint of water.

Our skin is the region in which the electricity of the body naturally accumulates, as its nature is to collect at the terminations and on the surfaces of conductors. (Faraday had a chamber constructed of a hemispherical shape of wood lined with tin; he placed himself in the middle, while powerful electrical discharges were made into the tin from without by his assistants, and as he had foretold, he could not feel any effect of the electricity whatever, as it all clung to the surface.) Nevertheless whenever inflammation has set in in the epithelium the natural state of affairs has become altered to some extent. It is true that electricity as before obeys the law of operating on the surface, but the operating, attracting and excited surface has in this case become the inner serous coating. This is especially the result of colds, since the nerves require a certain amount of heat to enable them to functionate electrically and thereby to attract the blood. If this necessary degree of warmth is withdrawn from the superficial nerves of the skin by changes of the weather, which is frequently the cause of inflammation, the blood passes to the interior surface, where the nerves still retain their power of attraction: but as in this case the considerable quantities of blood which were formerly apportioned to the whole exterior surface are now crowded together in the small space of the interior serous membrane, stagnation, over-heating, and inflammation must necessarily result.

Inflammation is the term employed, partly since the increased temperature is directly noticed by the patient himself, partly because as in the case of scarlet fever where owing to the fact that the nerves of the epithelium are paralyzed, the congestion of blood takes place on the outer surface and the rise of temperature may be measured directly with the thermometer: but the peculiar nature of inflammation is to be found in the coagulation of the fibrin which is the result of the over-heating of the blood and which has the effect that the capillaries in which this takes place become narrowed, and so oppose a resistance to the free passage of the blood, from which further disturbances follow.

It is thus plain that in the first commencement of such inflammations *a proper redistribution of the blood* which may in the case of colds be effected in the way already described by rubbing with vinegar or salt-water, will put a stop to the further spread of the disease. Packing in wet cloths too as it increases the electricity through difference in temperature, and so tends to redistribute the blood equally throughout the system may produce beneficial results. Both methods of treatment relieve the congested region of its excess of blood and the coagulated fibrine is granted time by taking up water to be reconverted into albumen which is then carried along in the circulation and put to use in the organism. This solution of the fibrine is facilitated by water containing Glauber Salts (Physiological Salt-water) which is of great value in all cases of inflammation.

Mistaken, and in many cases hurtful, yea, even fatal is the treatment so often employed in inflammations of applying *ice*. It is true that this appears to fulfill the requirements of Prof. Nothnagel: *"rest, cold, and local removal of blood"* but it only appears to do so. For what a great difference is there between drawing the superabundant blood to another place and the action of icy cold, paralyzing the nerves. Owing to the attempt to diminish the heat by ice, the congested blood is by no means drawn into the general circulation, but merely, owing to the mechanical contraction of the capillaries through cold, driven into the capillaries at a little distance from where the ice is placed; while the cold operates and the paralyzing action of the ice on the nerves is continued, no evil results are perceived, but as soon as the ice is removed the coagulated blood begins to decompose and poisons the rest of the blood in the body. Then we read in the papers:—"In spite of the fact that ice-compresses were applied with frequency, the disease was nevertheless not overcome and all the skill(?) of the phy-

sicians proved in vain". The proper way of expressing this would be:—
"The inflammation of the brain (or of the lungs) passed into putrefaction because the pedant of a doctor made the mistake of keeping
the inflamed blood fixed in one place by the application of ice, instead
of conducting it towards the skin and so distributing it throughout
the system."

As regards the third factor in the cure of inflammations,—rest,
it is certain that real rest, namely desisting from bodily work and keeping away emotional impressions, allows the blood to flow normally, and
so to dissolve and carry away any prejudicial deposits; but I am of
opinion, that the expression "rest" has another and a more valuable
meaning, i. e. on account of the horizontal position of the outstretched
body we ought to say instead of "rest", "going to bed."

"Going to bed" means keeping the surface of the body everywhere
at an equable temperature, which is a matter of much importance for
effecting the equable distribution of the blood, and in cases of a cold
may alone suffice to remove evil effects, especially when combined with
hot herb-tea. The sooner a man with a cold is put to bed the sooner
he will get well.

To this may be added, that in the case of the inflammation of the
lungs, the fibrine which has been deposited in the fine capillaries is
more easily dissolved away in a horizontal position, as the heart in
order to reach the apex of the lungs is not then obliged to overcome
the pressure of the vertical column of blood, which in an upright position corresponds to the distance from the heart to the collar-bone.
Thanks to the solvent capacity of the blood-serum, especially when
this has its normal contents of salt, an inflammation of the lungs may
be removed in 7—9 days, if the patient be put to bed sufficiently early,
without the employment of any other remedies. This is also the usual
case with individuals who have a normal state of blood and a robust
constitution, as the renewed blood revivifies the terminations of the
nerves, and enables them to eject the dead terminations changed into
mucus with the assistance of compressed air through the bronchial tubes;
so that a certain amount of rattling in the throat towards the end of
the eighth or ninth day is by no means a sign of aggravation but in
reality the beginning of recovery.

That in the case of older people whose blood has gradually lost
part of its sulphur, iron, and lime and consequently is diminished in
its power of absorbing oxygen, and from loss of salts in the blood has
lost part of its electrizing capacity, the result is more often fatal, is
a matter of common experience. Perhaps a change may be effected
in this respect when the employment of Physiological Salt-water owing
to the solvent action of Glauber Salts becomes more general also with
older persons.

Epilepsy. I have already given my theory of epilepsy on pp. 100

to 103, where I show that it is in the main due to disturbances in the circulation, and that these disturbances are in great part due in their turn to an insufficiency of oxygen in the blood. As when the supply of air is cut off from the fire in a boiler the water it contains cannot be made to boil and produce steam, the regular production of which forms the source of energy, so when an insufficiency of oxygen is inhaled the constancy of the nerve functions which by the attraction of oxygenated blood and the rejection of blood filled with carbonic acid put into activity the circle of motion is more or less prejudicially affected. From all that has been said it becomes plain, that this is the source of all functional derangement (see pp. 33, 34) and epilepsy only in so far calls for especial consideration, as the vegetative part of the nervous system (the sympatheticus) takes possession as it were of all the blood for itself to the prejudice of the cerebro-spinal system. Thus it is that the treatment of epilepsy is as it were naturally indicated.

The first endeavor should be to restore by means of the administration of sulphur, lime, and iron, the proper number of red corpuscles (v. p. 110) to the blood. The formation of blood is also assisted by the consumption twice a day of half a pint of boiled milk to which a little salt has been added; a certain amount of amorphous silica should (as mentioned on p. 111) also be administered so as to preserve the orderly connection between the blood and the nerve substance (see p. 15). While in this way the blood and the nerves are restored to their normal constitution, some plan must also be devised for causing the blood to continue to circulate regularly. This is effected by rubbing the body with salt and water (thirty eight grains of salt—a tea-spoonful—to a quart of water) both on getting up in the morning and on going to bed at night. Custom rules our organism so tyrannically, and our bodily functions go on so mechanically, that after these rubbings have been continued for a few weeks the circulation takes the orderly course of itself. It is nevertheless advisable to continue the rubbings a couple of times weekly to prevent relapses; for it stands to reason that as the habits of our body are so mechanical, even an abnormality which has prevailed for a number of years is only too liable to return when the new habit has only been introduced for a short length of time. Hence it is that during the action of new curative influences e. g. the electrical soil of Carlsbad, everything appears to be going on well, but after a return to the former way of living the symptoms of disease soon reappear. Hence in the treatment of epileptics care must be taken to keep the constitution of the blood and its circulation normal, even months after an apparent cure has been effected.

This latter end (a vigorous circulation) should be aimed at in all chronic affections not merely by rubbing with salt-water but in addition systematic and powerful inhalation and exhalation should be resorted to, particularly the latter. For this purposes I have recommended to my patients a little trick which is much more efficacious than most of the kinds of parlor gymnastics, and proves effective with all who are subject to convulsions. In Berlin it is widely known as Hensel's method of breathing. It consists in placing oneself behind a chair the back of which is grasped in the hands whereupon the "knee-bend" is made, the breath being expelled at the same time. After this has been done the upright position is resumed, the inspiration following of itself. It will be then observed that instead of the ordinary indolent respiration a very vigorous respiratory movement takes place, very much more air being expelled than was inhaled. This is explained by the fact that not only the inhaled nitrogen is expelled but also a certain amount of the carbonic acid gas that has accumulated in the venous blood and especially in the venous blood of the abdominal viscera. The cause of this is that the mechanical pressure due to the contraction of the abdominal cavity by the knee-bend, causes the elastic intestines to exercise a considerable pressure on the walls of the large veins. These walls are pressed together after the manner of an india-rubber tube with the result that their contents are forced upwards through the portal vein into the ascending vena cava, thence to the right auricle of the heart and thence to the lungs where the carbonic acid gas contained in it is given off. Every one who tries this gymnastic breathing exercise practically, will find that after every bend of the knees 3 inspirations and 3 expirations of unusual intensity follow each other; it is only then that breathing returns to its normal placidity, whereupon another bend of the knees may be made with the same result and then a third, and the exercise finished. This has occupied two minutes and in that time the whole of the blood of the abdomen has been completely renewed, new oxygenated blood having replaced what was there before. This is due to the nature of the circulation in which the blood from the heart in accordance with the law of gravitation enters the abdominal aorta passing thence through the branch-arteries to the stomach, liver, spleen, kidneys and the other viscera, whence having given up its oxygen it enters the portal vein, where when the respiration is deficient it remains for a greater length of time than is desirable from a physiological point of view. The longer it remains in contact with the organs of the abdomen the more completely it gives up the oxygen it has brought with it from the lungs, and is loaded more and more with the carbonic acid gas produced by oxidation, which diminishes the activity of the nerves and especially affects the movement of the diaphragm, to which fact are due the stoppages in the domain of the portal vein which appear at times as cramp of the stomach, at times as colic of the intestines or of the gall-bladder, in other cases as hysterical convulsions, and when the mischief has gone so far that nearly the whole of the blood collects in the roomy veins of the abdomen, then we get those disturbances of the cerebro-spinal system that have received the name of epilepsy.

We thus see that care must be taken to prevent the overfilling of the portal system of veins with venous blood, and the method of respiration above described should be repeated every two or three hours, which is often sufficient of itself to prevent the recurrence of the epileptic attacks. But as these exercises can only be performed in the day and when awake, and as is well known the epileptic fits often occur at night or just before waking up in the morning, we must attribute the attack to *the overfilling of the portal system with stagnating blood.* In regard to this latter point care must be taken in epilepsy more than in other diseases to avoid the "fattening-cure" which consists of drinking daily about two quarts of milk, which is by far too much. For every quart of milk contains 2315 grains of caseïn, butter fats and milk-sugar with about 92 to 108 grains of ashes, so that one quart of milk per day should suffice for the nutrition of the body; but taking into consideration that the nerves of the intestines in order to functionate properly require a certain amount of change (*"toujours perdrix?"*) it follows that also other nutriments should be supplied, so that one pint of milk in the course of the day in addition to the rest of the diet is fully sufficient. An excess in this direction tends to promote the attacks of epilepsy. In fact all epileptics should be earnestly warned not only against excessive milk drinking but against an excess in diet of any kind and especially against heavy suppers. A plate of milk-soup and a piece of bread and butter are fully enough supper for any epileptic patient. Disregard of this rule is frequently followed by convulsions during the night or in the morning. The explanation is to be found in the fact that the blood at night always tends to flow into the abdomen in order to supply the sympathetic system with the necessary oxygen to enable it to carry on the processes of new formation (absorption and assimilation, digestion and secretion). If now the intestinal canal is loaded down with twice as much food as is necessary for recruiting the bodily waste, twice as much blood will collect in the abdomen for the purpose of carrying on the processes of digestion. From this it follows however, that the brain is not only deprived of that amount of blood of which it is usually deprived during sleep, but even of that amount which is indispensible to produce the renewal of the lymphatic juice which should nourish the substance of the brain. From the point where the depletion of blood from the brain exceeds a certain degree, the independent proper electricity of the nerves comes into play, which else is tied in a close circle through the electricity of the salts of the blood and through the circulation; within this circle the electric circuit returns into itself and is usually not perceived. It is first perceived in the moment when the wave of blood which forms the connecting link for closing the electric circuit of the terminations of the nerves, temporarily fails. Even the mere partaking of salt, because this draws to itself water from the contents of the lymph, and the salt-water thus produced increases the quantity of the blood, may ward off an epileptic attack. In this respect it is indifferent whether we use common cooking salt (chloride of Sodium) or the expensive Bromide of potassium. The real cure of epilepsy cannot be effected thereby, for we may easily see, that the salts consumed, and indeed bromide of potassium as a foreign substance even more energetically than the common salt which is more congenial to the blood, will be removed out of the circulation by the activity of the kidneys. In consonance with this, experience teaches us, that more frequent and larger doses of bromide of potassium must be swallowed, without at last curing the epilepsy. Very many patients have affirmed this to me, who had bought bromide of potassium by the pound from the well-known source which largely advertises it.

No, bromide of potassium is *not* able to cure epilepsy, but this can only be done by the restoration of a well-ordered *circulation* and a normal *constitution of the blood*. Does it not bear witness of a far extended narrowness in the views on physiology, when it is thought that blood may be fed with only *one* substance? Has not bromide of potassium the grievous disadvantage, since it must be again eliminated through the kidneys, that together with the removal of this salt at the same time sulphur, lime and iron are removed from the circulation with the increased urinary secretion in larger quantities than before, so that the exciting cause of epilepsy is systematically increased.—Surely viewed from this stand-point we must condemn the sale of bromide of potassium, engineered through a well-known unmeaning literary circular, at least as the work of a physiological ignoramus, and every physician should warn against it as emphatically as the prophet Isaiah (ch. LVIII, 1) commands.

In addition to guarding against overloading the alimentary canal with food, epileptic patients should guard as much as possible against *excitements* and against *taking cold*. Mental excitements draw the blood to the brain and hold it there for an undue time to the injury of an orderly circulation. So also by colds the blood is pushed away from the external skin, and so the proper closing of the electric chain is interrupted, which puts the nerves of sensation terminating in the skin in connection through the mucous membrane with the serous (sensitiva) membrane of the nerve-center.

Similar harm is done by *dancing* and *skating*, after which the patients frequently succumb to an attack. This may be explained by the centrifugal force, which in a circling body drives the blood to the periphery and draws it away from other regions of the body, so that also in such cases the orderly closing of the electric chain between the nerves and the blood suffers an interruption.

That irrational dancing may lead even to a spasm of the muscle of the heart, which terminates immediately in death, of this the newspapers report every winter various examples. Epileptic individuals should not take part in dancing even in a rational manner, until their constitution has been fundamentally strengthened through an orderly *constitution of their blood* and an orderly *circulation*.

Fat, unhealthy accumulation of, s. p. 110.

Fever. The states of the body which show either an increased or a lowered temperature of the body are together called fever. It is of characteristic significance, that the urine of fever patients is found poor in mineral salts, and that the administration of such salts lowers the high temperature of the blood. It seems therefore plausible to consider the *lack of salt* in the blood as a fomenting cause in certain fevers. Since with the loss of salts also a loss of iron occurs through urination, the diminished contents of iron and salts in the blood are to be considered the cause of many states of fever, the more so, as salts and iron cure a great many diseased states which accompany fever. These mutual relations appear even more manifestly, when we endeavor to fathom what processes actually take place in fever. If we first inquire whence *ague* comes and then whence the *high temperature* originates, we reach the following results:

A shaking chill or ague is the vibrating of the external or the internal membranes of the body, or rather of the nervous fibres terminating in the external or the internal membranes, caused by the intermission of the arterial wave of blood which should close the electric circuit. Whether this has been caused by a *cold*, which has caused the flowing blood to recede from the terminations of the nerves owing to the narrowing of the capillary vessels, or that an intense emotion has taken place, or that a positive lack of blood caused certain portions of the organism to be deprived of arterial blood, e. g. the external skin or the mucous membrane of the intestines, or the spleen, or the liver,—in every case a chill or shaking ague, thus a sensation of *cold* is the first stage; only after a certain interval follows the sensation of *heat*.

Now what can be the origin of the cold?—The answer is, the combustion i. e. the oxidation of nerve-fat has ceased in the respective region from lack of oxygenated blood. Without combustion no warmth.

But whence the heat, which follows? Is there effected afterwards an increased consumption of nerve-fat?—No, this is not the case. The heat has quite another source. And what?—The same cause is operative, which warms the contents of the mashing-vat, when the grape-sugar separates into alcohol and carbonic acid, while the liberated force which conserved alcohol and carbonic acid in the form of sugar, is now perceived as converse electricity, or electricity operating in a changed direction, thus as *warmth*. Our blood also is heated, when after previous partial stagnation and cessation a fermenting disintegration of the albumen of the blood takes place through the detachment of ammonia. The chemical nature of the products of disintegration of the albumen of the blood may indeed be very various, according to the organ chiefly involved. In the pancreatic gland e. g. other products of the disintegration of albumen are evolved than in the salivary gland of the jaws, the tongue or the ear, others again in the spleen, the liver, the stomach and in the intestines. As in the *normal* so in the *diseased* state. The processes of disintegration take a devious course, but always so that in the duodenum a certain kind of products of disintegration is found which does not occur in the liver, nor in the stomach, nor in the spleen; and so again we find products of decomposition of the albumen in the spleen which are characteristic for this gland and which would be looked for in vain in other organs. At present these chemical products of the disintegration of albumen are yet with much insistence characterized by many physicians e. g. by Dr. W. Albert Haupt, as organisms standing between the fungi and the algae, and denominated on the one hand *bacilli* if they have the shape of little rods, on the other hand *cocci* if they have a globular form. From the fact that certain *combinations of aniline* have the form of bacilli and others that of cocci, taken in connection with the chemical fact, that also the coloring matters of the gall are combinations of aniline, it follows that all bacilli and cocci belong into the general category of disintegrations of albumen allied to aniline, and of these each gland produces its specific kind.

If chemical disintegrations of a special kind thus take place, the disease of fever takes on itself a specific character, so to say a well-defined coloring, which causes it then to be denominated scarlet-fever, gastric fever, typhous fever, hectic fever etc.

In the urine there are always found certain elemental parts of the substance of the blood or of the flesh, especially urea and ammonium-urea, and these give a clear proof, that the cause of the fever-heat is found in the electrolytic disintegration of albuminous substances.

But this knowledge also gives us the means of putting a stop to the fermenting disintegration, and of healing with the fever at the same time the whole disease.

A universal remedy in this sense is vinegar. As this protects meat in the pantry from decay and putrefaction, because it enters into a chemical combination with the gelatine-sugar of the albumen, so also in the living body. Sugar-water mixed with vinegar is a remedy to be recommended in every kind of fever; it guarantees at once by its cooling effect, that the processes of fermentation which presuppose a certain amount of heat, are stopped. Lemon-juice and hydro-chloric acid have a similar effect. Five drops of hydrochloric acid in a tumbler full of sugar-water are sufficient.

Still more decisively, however, is the electricity which is about to disappear, conserved and brought into renewed activity by means of Physiological Salt-water, in which sulphates, phosphates and chlorates form an electric chain of potentized activity. Give to the fever patient physiological salt-water, in which 120 grains of the salts are dissolved in one quart of water. It satisfies the thirst, moderates the heat, electrizes the intestines, the walls of the arteries and veins and the kidneys, and stops the fermenting disintegration of the albumen of the blood. The essential point to be observed is the dose, 120 grains or 6 scruples to the quart of water. Such a proportion corresponds in a certain degree to the contents of salt in the human blood; therefore it also operates as blood, giving vital power, without introducing foreign substances into the blood, as is done by the administering of quinine, strychnine, arsenic, antipyrine etc.

Vinegar and water or salt-water as a beverage are of course not sufficient to produce a cure, when diseases like disintegration of the blood are to be removed, which appear accompanied by fever e. g. in malarial fever and yellow fever. In these affections we must supply a substitute for the hemoglobin which absorbs oxygen and which has been destroyed in considerable quantities by the ammoniacal and cyanic products of the disintegration of the stagnating blood. This substitute is afforded in physiologically correct manner by the formiate acetate of the sesquioxide of iron and calcium (see pp. 135, 136, 141), i. e. Hensel's Tonicum.

Already in the year 1879 I cured in Newark, N. J. numerous cases of malaria and intermittent fever in which quinine had been ineffectual, by means of this preparation in an astonishingly short time. I myself made use of the tonic limonade at that time to preserve my health, and this with such success, that I remained able to work, while many others sank down wearied and nerveless. Since that time this preparation has acquired a great repute as a prophylactic against prostration in tropical regions. Of the numerous proofs of its restoring and preserving power against tropical influences, I would here present only one, because it comes from an observer of peculiarly clear judgment. The notice is found in the "Mittheilungen der Ostschweizerischen Geogr. commerc. Gesellschaft in St. Gallen 1 Heft 1888" and its author is a certain Mr. C. Stolz. In an article entitled "Das Leben des Europäers in den Tropenländern" he warns against the use of unboiled water as causing fever, especially in the forest regions. Instead of it he recommends the use of unfermented palm-wine and closes with the following remarks:

"I would like to direct attention to another beverage, namely the "Tonicum" invented by the chemist Hensel. I have convinced myself theoretically by reading Hensel's writings and practically through the use of Hensel's Tonicum, that this easily digested and pleasant tasting preparation of iron acts most beneficially on the organs most apt to suffer, namely the spleen and the liver, and is a most valuable beverage for tropical regions. Without the disadvantages of wine or beer, it is more refreshing

than any beverage known to me, and is also valuable for the reason, that it will serve
to disinfect doubtful water, a great desideratum in traveling. It is desirable that all
who journey to trans-oceanic countries should acquaint themselves with this remedy,
which would prove a frequent relief."

The author then mentions the swelling of the spleen and the liver
which accompany bilious vomiting, and which he endeavours to explain
in his way. He did not recognize the cause of this ailment, viz. the
overloading of the blood with carbonic acid, but he correctly remarks,
that the Carlsbad-salts deserve the preference over calomel and aloes,
which are used by Englishmen, and he adds, but above all is to be
preferred Hensel's Tonicum which produces the normal secretion of
bile and frees the liver and the spleen.

To this I would only add, that practical observation, even without
clearly comprehending the physiological causes, has discovered the truth,
namely that *no one* substance will of itself avail, but there should be
a rational change between tonic lemonade and physiological salt-water,
the natural instinct being herein the safest guide. If the physiological
salts are not within reach, the artificial Carlsbad salts may be used
without any misgiving, because this also contains sulphates and chlorates together and operates not only by electrizing, but also protects the
blood from putrefying disintegration, while the tonic lemonade enables
the blood, after red blood-corpuscles have been formed, to chemically
hold the oxygen respired, thus securing the normal function of the abdominal nerves, and to keep up the circulation in the diaphragm,
through which the blood is freed from the carbonic acid which paralyzes the nerves. These two beverages, as they contribute to keep
liver, spleen, stomach and intestines capable of normal activity, at the
same time favor the orderly function of the kidneys, with the result
that the urea eliminated by the oxidizing respiration is quickly removed
from the circulation, before it becomes by longer stay in the blood-
channels carbonate of ammonia which lames the power of the nerves
and poisons the blood.

If we now put together the protecting agencies against climatic
fever we find the following results:

1. Counteracting the internal overheating and the disintegration
of the blood through cooling beverages, which keep up the functions
of the kidneys;

2. Emancipating ourselves in part from the damp atmospheres
which conduct away electricity, by the internal generation of proper
electricity (motion and respiration);

3. Supplying the blood with the material by which it is enabled to
absorb the nerve-quickening oxygen;

4. As soon as ammonia appears from the stagnation of blood in the
spleen or the mesenteric vein, which manifests itself by a shaking chill,
the ammonia is to be rendered innocuous through acid drinks, and thus

the progressive chemical disintegration of the albumen of the blood is to be quickly stopped.

These four conditions are satisfied in a rational manner by physiological salt-water and tonic lemonade. In the mean time other factors may also have their prophylactic uses. In this respect I have defended in private conversations a theory which I have not hitherto made public; but I shall not now keep it back any longer, especially since also Mr. Stolz in his above mentioned article has touched upon the same subject. He states the following concerning the subject of Cholera:

"As coppersmiths are never seized with cholera, wearing a thin copper-plate upon the skin has been recommended as a preservative against this disease. It might appear an assumption for a layman to enter on such details. I do it simply in order to encourage others, to make their own observations, and when necessary to act independently."

It can in no way be denied that the very touch of metals has an electrizing effect. In so far as our skin encloses a salty fluid, and is formed of carbo-hydrates and combinations of cyanogen, the regions of the body which are further distant from one another may by metallic conductors be brought into relation with one another and be enabled to functionate more energetically, even if the metallic covering should not furnish the conditions for an independent electric circuit. Perhaps it may thus be explained how even the much abused "rheumatism-chains" had their use, although the attempt has been made, to "scientifically" deny this. Metals have to be considered as concentrated sources of electricity. Has not partial paralysis been temporarily removed by the laying on of magnets, thus through the laying on of a metallic substance? The armor of the Middle Ages was no doubt also of a certain physiological significance in that it augmented the proper electricity of the body and in such a way contributed unbeknown to heighten strength and courage.

Now since metals at the same time have a *cooling* influence, as may be manifestly perceived, when we press a knife-blade against the forehead, and since protection against overheating is in tropical regions invaluable, it would be worth while to try, whether the wearing of a fine nickel-plated shirt of mail would not increase man's power of resistance against tropical heat.

Abstinence from alcoholic beverages, because these deprive the blood of oxygen, and abstaining from emotions and excitements because these predispose to stagnation of the blood in the region of the portal vein, as a rule of hygiene for the tropical regions, should not require any special mention. Another maxim of experience is that we must not make our home in close proximity to water, if we would not subject ourselves to fever.

With respect to food, the vegetarian diet is decidedly to be preferred to the carnivorous, as long as the hot season lasts; on the other hand a warming meat-diet during the rainy season is said to prove preservative of health and even essential in recovery from fever, which may in part be explained from the fact that lean meat contains iron, as 100 grains of meat contain .04 grains of iron, and thus meat supplies the material for new hemoglobin.

After our explanation on page 120 concerning the results of Hensel's Tonicum, and our explication of the necessity of the salty serum of the blood for the sake of preserving the electricity of the body, it is to be hoped that physiological salt-water and tonic lemonade will in future put an end to the devastation of yellow fever in the capital of Brazil, which has usually raged there among late arrivals from Europe during the summer months of January to May. A necessary condition would, however, have to be, that another hospital would have to be erected on the mountains surrounding the Bay of Rio Janeiro instead of the hospital San Sebastian. The journal "Paiz" which appears in Rio in the Portuguese language brings in its issue of April 23d 1891 in the article "The Ante-chamber of Death" a remarkable communication concerning the revolting condition of things which obtain in the fever-hospital of the Capital owing to its unfavorable situation. Considering the increasing immigration to Brazil of subjects of the German Empire, the German Government should cause the article that appeared in the "Paiz" to be published as a warning as widely as possible.

The article shows that the hospital San Sebastian has the most unfavorable situation imaginable in sanitary respects. It lies in the most unhealthy portion of the district of Retiro Sandoso, where malarial fever always reigns, close to the sea-shore. The strip of land separating it from the sea is low, flat and exposed to inundations. On the one side there is an extensive tannery, the exhalations from which are really unbearable. At a little distance, in the middle of the bay lies the island Sabucaia. The suffocating smoke of the great fires in which the offal is burnt on this island is daily wafted over to the hospital. The remains of old rags, dead animals, spoiled fruit etc. which are daily cast into the sea from this island are floated all the way to the hospital. Behind the hospital there extends an immense Manga, a swamp and cemeteries. In short the situation of the hospital is as if it were selected to attract the most deadly malarial fevers. No wonder that of the officials at the hospital one third is always sick.

No one can have any conception of the filth and disorder in the sick chambers. The sick are disrobed soon after their arrival and clothed with the hospital shirts. The clothes worn, dirty and filthy as they may be, are pushed under the bed. *It is hours before these clothes reach the disinfecting station.* As there are no slippers in the hospital, the dirty socks and shoes of the sick are left beside their beds. By their side are standing iron water-cans, spittoons, medicine flasks, urinals and between them dirty blood-stained linen cloths. All these spread an unbearable odor and assault the life of the unfortunates who are compelled to breathe this air. The mortality among the sick brought here is said to be nearly one half, those who died during the transportation not being included in the estimate.

All the sick received have administered to them a strong dose of castor-oil; in their second stage they receive a drink containing iodide of iron, and as the last medicine strong coffee. The use of physiological salt-water as a substitute for the electrizing serum is there as yet unknown.

Hemorrhoids. Take every morning immediately before breakfast a dose of flowers of sulphur; at noon before dinner a dose of hematite; in the evening before retiring half a tea-spoonful of nerve-salt dissolved in a wine-glass full of water. Thrice a day methodic expiration by means of gymnastics is to be obtained, and every morning the breast, back, abdomen, arms and thighs are to be gently rubbed with salt-water (see p. 145). Drinking beer or wine, smoking and a sitting occupation are injurious.

Cutaneous Eruption.—Elephantiasis.—Leprosy.—Scrofula. Cutaneous diseases and glandular swellings arise in consequence of the deterioration of the lymphatic fluid, when this does not through nutrition obtain a sufficient quantity of lime and sulphur to keep together the albumen. One of the most curious forms of cutaneous disease is elephantiasis, so called, because the lower part of the thigh, the subcutaneous connective tissue of which is interpenetrated by numerous lymphatic vessels, swells up so immoderately, that it becomes like the clumsy shape of an elephant's foot. The history of ancient times has brought several examples of this to our knowledge. The cases involved were always gluttons who indulged in a gross meat-diet and an abundance of wine. Now since in meat and wine only traces of combinations of lime and of sulphates are contained, while phosphate of potassa is found in it in abundance, (only 1 part of sulphates is found to 70 of the phosphate of potassa, while blood of normal constitution contains about twice the sulphates that it contains of phosphates,) the deficit in sulphur in the blood and the lymphatic juice is thus sufficiently explained by the insufficient supply of sulphurous aliments for a number of years; thence it is that the nature of cutaneous eruptions is easily studied as to its principles in the elephantiasis of certain gluttons of antiquity.

In these latter days I had the opportunity once of studying elephantiasis in the case of a beggar. As I approached him a strong scent of alcohol met me. He was therefore a drinker (*potator*), and in consequence also in this case the lack of lime and sulphur in the blood and lymphatic juice was recognizable as the cause of the disease, for alcohol contains no sulphur, and habitual drinkers are not accustomed to consume much solid food; nevertheless the respiration continues, and the products of the oxidation of the albumen of the blood which contains sulphur goes on, leading to the formation of sulphates which are eliminated in the urine more and more every day. Under these circumstances the phosphates of the flesh (phosphate of creatin), of the nerves and the bones gain eventually the exclusive sovereignty. No wonder if the material of the bones full of phosphorus expands like a sponge and the phosphoric material of the cutaneous nerves turns into maggots and worms (Pedicular disease). This process is to be compared to the formation of the fungus in houses (*Polyporus destructor*), which grows out of wood which by long continued moisture has been deprived of the easily soluble sulphates. Without sulphur no Protein. In the wood-fungus the coherence of the cellulose of the wood is lost, and instead of it is found a new and peculiar formation of tissue, which is based almost exclusively on phosphate of potassa, and imitates in a striking manner the ramifications of the sympathetic nervous plexus which is based on ammonium phosphate.

In a similar manner by metamorphosis there arise from the bark of trees which predominantly contains phosphates, lichens and mosses, which also show no sulphur in their ashes. And just in the same way herpetic eruptions take place on the human skin, when their lymphatics show a protein deficient in sulphur. Even the well-known cutaneous disease which we call syphilis, most easily finds an abiding place in individuals of a weakly constitution, who have not sufficient lime and sulphur in the blood and in the lymphatic juice, to offer resistance to the infectious disintegration; and this latter is cured by the mineral water at Aix-la-Chapelle which contains sulphide of sodium, sulphate and carbonate of lime.

Leprosy falls in the same category. About a quarter of a million of the inhabitants of India are suffering from this affliction. This peculiar dyscrasy was the subject of a meeting called on June 17th, 1889, at which the Prince of Wales presided, and which discussed the impotence of the medicinal art in battling with this affliction. With external remedies it is indeed impossible to cure it, nor can it be studied in the single individual. We must rather view by the bird-perspective mode, the *manner of nutrition* in the various countries attacked by it. From this point of view we learn the following:

On the Island Hawaii the inhabitants chiefly live on sweet fruits. Bananas, Cocoa-nuts, dates, bread-fruit and mangoes, as well as the papaws with their melon-like fruits clustering around the stem, furnish the national food, to which are added fishes. The bananas contain an extraordinarily small amount of ashes, especially little of the sulphates, little potash and little lime. Bananas owing to their lack of earthy material are subject to decay shortly after they ripen. Fishes also contain but a small quantity of sulphates or of earthy materials (potash, lime, soda, magnesia, silica and iron); fish putrify within 24 hours.

The *Norwegian* lepers are recruited from such shore-dwellers and islanders as live exclusively on fish. Since the Norwegian lepers have been taken into hospitals where a proper system of nutrition rules, their number has fallen from 2000 (about the year 1867) to less than 700. In the English colonies on the other hand their number is steadily increasing. The reason of this is, that the Hindoos chiefly live on rice which contains almost only starch and but few earthy materials.

Also in Germany leprosy is found with the poorer fishermen and islanders on the Baltic Sea, who chiefly live on fish. In Italy it is the *"Frutti di Mare"*, and in France the one-sided living on sardines which predispose to leprosy. Year after year the sulphates, lime and iron pass away in the urine without any corresponding supply. This finally causes the chemical disintegration of the albumen in the blood, lymph and muscles and even of the bones which need earthy material for cement.

In *Germany* the plague, black-death and leprosy have ceased to be endemic since potatoes are universally raised and eaten. This is owing to the earths contained in potatoes. Dry slices of potatoes on burning leave behind 4 per cent of ashes, of which fully 10 per cent are composed of sulphate of magnesia. This may also explain the disappearance of leprosy from England since enormous quantities of potatoes are imported there from Germany.

The Irish, because they eat herings and potatoes are free from leprosy.

Good potatoes with an abundance of earthy constituents have no scurfy skin and can be kept all the winter, while potatoes manured with stable-dung are scurfy and rot easily. It is the earths always which impart durability and coherence to the meat of fruits, as also to the flesh of the muscles.

The close resemblance of the ashes of potatoes to the ashes of beef may appear from the following comparison:

Ashes of Beef.		Ashes of Potatoes.
Potash	41	58
soda	$2\frac{1}{2}$	3
lime	$1\frac{3}{4}$	3
magnesia	$3\frac{1}{3}$	5
hydrochloric acid	5	3
sulphuric acid	$3\frac{1}{3}$	6
phosphoric acid	$34\frac{1}{3}$	16
silicic acid	2	2
carbonic acid	7	4

Taking all things into consideration, leprosy it to be regarded as an intensive scrofula, originating in defective mixture of the blood.

Like as children become scrofulous, if they receive watered milk without
the addition of salt, so leprosy is developed in India, where the government has imposed so heavy a tax on salt, that it can not be paid by
the poorer classes. Salt secures meat against putrefaction, against
chemical disintegration, because it electrizes. If the duty on salt in
India should cease, leprosy would cease. For the sweet potatoes
(batatas) which yield about two percent of ashes and are largely raised
in India form a suitable complementary aliment with rice. A sufficiency of all the earthy constituents of the blood must be at its disposal. Among these is *iron* of which the ashes of beef contain about
1 per cent. But since the use of meat is injurious in the tropical
regions, as the bases of meat xanthin, sarkin etc. (see p. 67) are
disposed to separate into *prussic acid* and *ammonia*, if a sufficiency of
oxygen is not inhaled, as is the case in the thin hot tropical atmosphere, it is advisable to take daily 10 to 15 drops of Hensel's Tonicum
on sugar. Besides this, Physiological Salt-water, which operates owing
to its contents of sulphate of potash and soda, has a regulating effect
on the normal constitution of blood and lymph, after a firm foundation
for normal albumen has been laid by the use for a time of flowers of
sulphur and of Calcium-Magnesium phosphate, whereby eruptions disappear. The same treatment also proves successful against herpetic
eruptions of every kind and also against scrofula.

Consumption. Even before Herr Koch and his adherents had
accused the "bacilli" of being the cause of consumption, the same false
course had been followed, even before bacilli were discovered. Men had
hoped to discover the essence of consumption, by examining all parts
of the respiratory apparatus with the microscope. The dead mucous
membrane was stretched on cork, it was dried, cut into slices, the
slices were then again softened in vinegar, and then the malefactor
was sought for with the microscope.

The sputa also were examined as to their specific weight, color,
smell, transparency, toughness, fluidity and their swimming on water;
it was also examined microscopically with the hope of discovering something peculiar between the epithelium, fibers, fat and the little bodies
of pus.

The expectoration was also examined chemically as to its contents
of phosphoric acid, sulphuric acid, chlorine, sodium, potash and silica,
in order to discover if possible the cause of the putrefaction of the
secretion; the hope was especially entertained of discovering organisms
causing fermentation after the manner of the fungus of yeast. No importance was given to the absence of sulphur.

Also the individuality of the muscles, which yet form only an
appendix to the nervous substance, e. g. the weakness of the *transversus abdominalis* for lifting the ribs has been seriously considered.

The thorax was measured, the intercostal spaces were inspected, the bounding lines of the lungs and of the heart were found out by percussion, also the especial "sound-color" of the lungs themselves, so also the sounds of the respiration were determined in some cases as metallic, in other cases as being of a peculiar timbre resembling the sound of a pasteboard-box etc.

The force of the expiration was measured by the spirometer.

It was thought of sufficient importance to examine the kind of fremitus and the sounds of the heart and strictly to distinguish whether a sound at the contraction of the heart emanates from the mitral valve or the tricuspid valve. The increase of the respiration up to 40 a minute was derived from the difference of air-pressure against the inner and the outer walls of the thorax.

In order not to be inactive single *symptoms were attacked*. The pains were combatted with opium in the form of Dover's Powder or morphine, water of bitter almonds and extract of belladonna; the expectoration was treated with lactucarium and hyoscyamus; the diarrhoea with sugar-of-lead.

The cause of the disease was thought to be, disturbance in the circulation of the venous system, combined with fatty degeneration of the heart, atrophy of the liver-cells, and amyloid degeneration of the kidneys; to this was added atrophy of the lung tissue and especially the decay of many capillaries of the lungs.

That the lack of oxidizing, electrizing oxygen in the blood might be the cause of the chemical disintegration of the tissue-substance, was not considered, because this lack could not be seen with the microscope, and the knowledge of oxygen gas as the aura of life and of respiration is not much older than 100 years.

Instead of this they distinguished the more 'precisely between "Phthisis acuta, typhoïdea, thoracica, asphyxitalis, catarrhalis, pneumonica, pleuralis, abdominalis, peritonealis, tuberculosa, granulosa, suppurans, pepsiniformis, clinica" etc. etc. For such "differentiation" is very meritorious, but "generalization" is one of the most stigmatized transgressions of a medical practitioner.

Did it need all this expenditure of acumen to realize what was here the matter? Was not one look at the pale-blue lips and the sulphur-yellow or grey cheeks, another look at the whole form, and finally the lack of resonance in the voice sufficient, to conclude from the cyanotic appearance and the dropsical symptoms the deficiency of the supply of oxygen, and thence the paralysis of the nerve of respiration and its appendages?

Every tissue which remains inactive is subject to chemical disintegration. It must degenerate and perish. Whether *cheesy* degeneration (tuberculosis) or fatty or suppurative degeneracy (catarrh), is pretty much indifferent, for it all amounts to mortification, and the decay of the material formed under the influence of the nerves cannot be limited to the epithelium of the lung-cells, but it transfers itself as well to the mucous as to the connective tissue, and as soon as the symphathetic nerve is seized, it is also transferred to the serous membrane.

In which way then is consumption acquired? let us put together a number of typical examples.

In close rooms, with closed doors and windows between the high rows of houses in narrow streets, especially on the damp ground-floor with sedentary mode of life, and contracted abdomen, especially with bad nutrition, with care, anxiety and destitution, men become consumptive.

Store-keepers who have few customers and little exercise, who never leave their store, and besides smoke tobacco fall a victim to consumption.

Seamstresses, who do not earn enough to heat their room in winter, and to fill their stomach, succumb to consumption.

Stone-masons, because they fill their lung-cells with dust of silica, instead of pure air are exposed to consumption.

Esquimaux who have been accustomed to breathe thick, heavy, cold air, when they have been exposed for show in warmer countries for a few months, without any exercise, die of consumption because they breathe nitrogen instead of oxygen.

Gorillas, chimpanzees, in general all monkeys who are penned up in heated rooms with diluted air, which rooms are neither aired nor supplied with fresh oxygen, while the animals are wrapped up in woolen blankets and deprived of any more extended exercise, become the victims of consumption.

Our daughters whom we prevent from taking part in the work of the household, and whom we keep from beneficial exercise, whom we deprive of the cheering enlivening enjoyment of a change in their surroundings, and of an interesting talk while doing the marketing of the family, of whom we would not require the healthy mounting of stairs, the bodily activity in the kitchen, the pantry and cellar, whom we deprive of the blessing of bodily work, with which we would rather favor the cooks, who acquire from it not only blooming cheeks, but also money in addition,—our daughters we sacrifice on the altar of consumption.

In the period of their development girls need more especially energetic bodily motion. Instead of this we continually hear remarks like the following: "She is too weak! We must not strain her, everything hurts her. The Doctor *also* says, we must spare her, she is yet too tender."—Yes, yes, spare her every exertion and as the proverb says: "What we save from our mouth, cat and dog will devour."

Labor! Work! Breathing! Motion!—That is all she needs.—If air enters her lungs, life will flow to the tips of all the nerves. Else the impulse to a further development of the body is lacking. The change of material is stopped, and since life and health rest on steady renewal, on the consumption and the growth of the bodily substance, the non-use and sparing of the body means its chemical disintegration; whatever does not advance, retrogrades.

Whatever we may choose to call it, atrophy or degeneration, it is indifferent, a *retrograde* process is taking place. Formed material is turned into unformed or half-formed material; sugar is turned into lactic acid, and lactic acid in the blood gives occasion to the disintegration of albumen, to the detachment of Leucin and Tyrosin, to putrefaction; but a putrefying secretion is poison. Whether it stagnates in the ovary or in the uterus; whether the deadly issue appears in the lungs or in the peritoneum, it is and remains *consumption*, i. e. a retrograde motion, disintegration of tissue. For it is never the lungs alone which are affected, for *all* the glands are degenerating, some faster, others slower, according to the *locus minoris resistentiae* (place of least resistance) and the hereditary or acquired neglect of the organs or the insults offered to them, e. g. lacing together the liver by lacing the corset.

What then does all this far-fetched differentiation in the varying image of Phthisis amount to?—*Nothing* at all, it only turns us from the goal, even as it appears to be merely a conventional makeshift to name diseases after the various organs. Even eye-diseases in very numerous cases amount only to dyscrasy, as may be seen with one look at the patient. As soon as the blood is improved, the eyes are cured in most cases in a surprisingly short time.

Dyscrasy means lack of electricity, just as *poverty* means lack of *money*. It would surely cause a smile if we should describe a poor man in the following manner: His coat is ragged, his knees stick out of his trousers, the heels are crooked on his shoes and the uppers have gaping holes; he has no socks at all, his hat is mashed, half of the brim is gone and in the crown there is a big hole. His pockets are quite empty and lacking not only in gold, but also in silver and copper coins. His hair hangs down unshorn from his head, his beard is long and uncouth, his cheeks are not shaved. His face is haggard and pale. —A single 200 dollar bill can cure all these defects. Just the same is the case in chlorosis in which so many symptoms appear: headache, flushing, asthma, difficulty of breathing, dizziness, loss of appetite, ringing in the ears, flickering before the eyes, disturbed menstruation. The same is the case in consumption. Your liver-trouble when rightly viewed is a kind of *consumption*, and if you walk about and breathe energetically in Carlsbad, your consumptive liver will be cured.

Many a one pities the mountaineers near Triest, who daily mount and descend the zigzag mountain roads to sell their victuals in the city; but these heavily laden people do not need our care, but those who sit still. The former get old and drink daily fresh oxygen.

As soon as *we* climb the mountains in Silesia or in Switzerland,

how painfully do the compressed air-cells of the lungs labor with many
in the outset; but with every hour the mountain-climbing becomes a
greater pleasure and we begin to blossom out. Our muscles become
more vigorous, the energy of our soul augments; we laugh and sing.
And yet nothing great has come to us, only a little fresh air.

On the other hand, when the diaphragm rests, breathing is no
breathing but sighing. The liver does not receive enough oxygen, to
produce the necessary higher temperature by oxidizing the nerve-substance, so that bile able to. assimilate food, may be secreted. Resting
on the cushion of the intestines, hanging suspended from the diaphragm,
the liver is too little pressed upon from either side to promptly forward
the venous blood into the Vena Cava, and the lungs do not suck it up
with sufficient vigor, therefore the liver works more wearily day by day,
while some regions lie desolate, suffer fatty degeneration. And all this
is caused simply by insufficient respiration. The lungs succumb to a
similar fate only modified by their loose tissue. Let us hear a consumptive patient, whom we will cite from her grave for the purpose tell
us her tale in her whispering voice devoid of resonance:

"The medical Councillor takes so much trouble with me and looks
after everything; he has examined me everywhere, and I am weighed
every two weeks. But my strength is too little. He has also forbidden
me to do anything or to leave the house; I am to do absolutely nothing. But nevertheless (!) I have to perspire so much. The less I do,
the more I perspire, If I read a little, I perspire; if I talk a little
with you, at once I perspire. I have such internal anxiety, as if I had
killed somebody. And I cannot eat anything at all. Best of all I.
should like to eat a pear, but the doctor thinks, that this is a faulty
appetite. Coffee I must not drink either, he thinks, it excites me too
much. He has ordered coffee made of lupines for me, but I loathe it
so that I cannot look at it any more".

These dark times, in which the theory of the physicians allowed
the patients to starve lies scarcely more than 20 years behind us. And
scarcely 10 years have passed, since a medical establishment in Baden-
Baden applied inhalations of *nitrogen* in place of oxygen, as being
advised by the highest medical authorities.

The patients perspired from internal anxiety, because the only
remedy, exercise in the open air was denied them; and if, following
instinct, they asked for malic acid, which is contained in pears and in
most fruits, and is able to neutralize the ammoniacal products of destruction in their tissues, it is called a faulty or depraved appetite.

Even at this day they send patients to warmer climes. As if
external warmth could help. The heat must be generated *within* by
the combustion of the bodily substance. Out of his own body the
pinions must grow with which the bird would fly; he can not fly with
feathers which are merely glued to it after his own have fallen out.

What one-sidedness further with those who would cure consumption with koumiss. It is true that the Calmucks who drink koumiss do not suffer from consumption, but they are indebted for this not to koumiss, but to the circumstance that they even as boys six years old, standing on the horse behind a 10 year old comrade, dash away over the sandy steppes, breathing their pure air free from dust, for the hoofs of the horses dash the dust out behind.

If the drinking of koumiss or kefir has a certain curative effect, this is owing to the contents of lactic acid, which is contained in this beverage gained by a peculiar fermentation of milk; for lactic acid is as useful for the stomach and for digestion, as it is hurtful when formed in the lungs. It neutralizes like malic acid the ammoniacal products of decomposition, and as the milk-sugar and the casein are already partly chemically split up, koumiss is an aliment that is easily digested. But common thick milk beaten up and mixed with half the quantity of Seltzer Water is of as much value as Kefir, and cannot be distinguished from it either in taste or effect.

From what has been said about breathing nitrogen instead of oxygen as being a fomenting cause of consumption, it may be easily seen, how we may guard against it. We must breathe vigorously, work vigorously and eat vigorously. Furthermore we must keep the skin open for the exhalation of carbonic acid, by the use of water and soap.

To this should be added the protection of suitable clothing to protect against injury from humid foggy air, which conducts away man's electricity. To breathe dry wintery air is not only not injurious but as we see in Davos even wholesome. Our dwelling-rooms should be dry and sunny; for *"Dove Sole non entra, Medico entra"* (Where the sun does not enter, the physician enters) the Italian proverb says. Then again we must think of cheerfulness of mind, by living in pleasant social intercourse.

Nothing so much protects against consumption as daily exercise in air free from dust. Happy he, who can make a choice of vocation in agreement with this maxim. Whoever is compelled to lead a sedentary occupation, should not miss in the free hours of the evening to put the air-cells of the lungs in activity, so long as the season in any way admits of it. That we must not exceed our strength, is a matter of course.

Suitable *clothing*, especially in a changeable temperature a woolen shirt, which prevents a sudden chilling of the skin, should be part of his A-B-C. That a man who will not renew the external cuticle by washing, will have to yield up his internal (mucous) membrane, we see in catarrh.

That *nutrition* must be sufficient, without going to excess is a matter of course. A *change* in food nerves man's spiritual faculties to their best performance. Man *is* what he *eats*. A one-sided nutrition of gruel, of which Hufeland adduces some examples, may increase a vegetative existence, but not the enjoyment of life. Spiritual energy cannot be generated by it in any eminent degree. On the contrary, Hufeland describes some of his old men as mere automatons, which seems to us neither enviable nor worthy of imitation.

A glass of wine at the right time, ripe fruit in its season, a glass of beer at the right hour, quicken the respiration and protect against consumption.

Not least in importance for securing proper undiminished respiration is the avoidance of annoyances and of disagreeable impressions.

»La joie est bonne à mille choses, mais le chagrin n'est bon à rien«
(Corneille).
(Joy is good for a thousand things, but vexation is good for naught).

Spiritual activity also, which, indeed, is almost always combined with bodily energy keeps us in good respiration. It is Psyche, who keeps the body young. That is seen in the learned world, the highest age has been reached by natural philosophers, whose occupation presents what is interesting, but nothing that is exciting.

Men not learned may enjoy the same advantages by taking part in public affairs. They are thereby compelled to their own advantage to leave their homes, to enjoy the open air, and especially to partake of the enjoyments imparted by sociability.

In company those who are predisposed to anxiety lose their oppressed feelings which prey on them in solitude. They breathe freely and inhale health. The "city fathers" on the average arrive at a good old age, especially if they also drink a social glass of good wine. Therefore those who wish to be protected from consumption should take part in public affairs. Women who personally preside over their household, do not need this; their spiritual energy springs from the lap of their occupation

Frauen-Arbeit ist behende
Aber sie nimmt nie ein Ende.
("Woman's work is nimble, But it has no end."

One thing is *needful for all: Activity!—*

When consumption has commenced, we can, now that we know our enemy, take steps to *help* and to *heal*, without regard to its stage, according to the following universal procedure.

The formation of lactic acid in the lung-tissue must be prevented, and the lactic acid already formed must be oxidized, and the current of blood, so long as the lungs do not resume their normal activity, must be *arterialized* by means of the *venous* system. For this purpose a goodly number of reliable remedies are at our disposal. These are substances which can give off oxygen when, having been absorbed by the chyle ducts of the intestines they reach the thoracic duct and thus

enter the venous systems and thus finally through the right part of
the heart enter the lungs. These substances are called *glucosides*, because they represent chemically condensed sugar. The mode in which
this condensation is effected, has been explained in the work "Das
Leben" (2 ed. p. 38), and I have also explained there (p. 47. 48.) how
from such chemically condensed sugar, fat is formed, while oxygen is
liberated. The first condition necessary for this change is indeed the
access of some oxygen, to start the process of transmutation; but only
8 atoms of oxygen are necessary to start this process, as a result of
which there are formed from 7 equivalents of sugar 42 atoms of oxygen
and one atomic group of tallow-fat. That is one of the great miracles,
which are at the disposal of the creative power of God, and it remains
a miracle although it may be chemically explained, for we cannot
imitate this miracle.

While we thus see an *oily fuel* as a new operative source of strength,
and at the same time *oxygen* under electric tension, because just liberated, arise from sugar, we already have in hand the most important means
for putting an end to consumption, which becomes visible even to a
gross vision, through the consumption of the fatty substances of the
body; and at the same time we make operative new vital force by means
of oxygen in electric tension.

New Vital Force!—Very little is needed for this. To make sure
of it we shall make the computation, for we must stand to our promise
of solving certain conundrums of life by means of simple numbers and
equations.

When we have hydrogen and oxygen gases in the proportion they
have in water, in gaseous form, an astonishing force may be liberated
by their chemical combination into water; since one kilogram (2 1/5 lbs.
avoirdupois) of such explosive gas in its chemical condensation exercises
such an intense hurling force that it would equal at the moment of
motion 23380 horse-power. This would make 23 horse-power for 1 gramme
(15.433 grains) of explosive gas, and as 1/9 in weight of this explosive
gas consists of hydrogen, these 23 horse-power depend on 1/9 gramme
(1.715 grains) of hydrogen. Thence it follows that one single horse-
power calls for but 4 milligrammes (.0624 grains) of hydrogen. And
it is indifferent whether the hydrogen consumed is in a free state or
in combination with carbon (as oil). In the latter case we have only
to use the 7 fold weight of substance, as the weight of carbon to hydrogen is as 6 to 1; so that we need 7 times 4 milligrammes (.4368 grains)
of oil to produce through its combustion one horse-power. From this
it follows, that *half a drop* of oil is sufficient for this purpose, if it is
consumed at the tips of our nerves through oxygen. Now when we
consider that our lungs need only to inhale 8 parts of oxygen in order
to liberate and set in chemical activity 42 parts of oxygen by the
oxidation of 7 united molecules of sugar, we are shown by the appear-

ance of such huge quantities of *vital air* in our circulating blood, air which we did not inhale, a source of power of which we had before this no knowledge. The propelling force of the heart, the pumping power of the lungs now appear in quite a new light. But we must, of course, always remember that such a result can only be derived from a coherent group of 7 molecules of sugar. The *single* molecule of sugar does *not* effect such a liberation of oxygen.

We must, therefore, ask: Where do we obtain such sevenfold sugar-molecules?—The answer is given in the work "Das Leben" (pp. 35, 45). As medicinal substance *Gum arabic*, as a physiological material the *gluten* of wheat deserve consideration. The latter arises from the glucose-group of starch through the chemical combination with ammonia, so that gluten (in contrast to the glucose formed into gum arabic by potash, lime and magnesia) may be regarded as ammoniacal gum arabic. In agreement with this, wheat-flour soups as remedy against consumption have been used among the common people for a long time already; so also gum arabic (Mucilago Gummi Arabici) is beneficial in all kinds of fever by its quietly oxidizing action. After the hitherto prevailing grossly empiric view, this effect is explained in such a way that the mucilage acts as an envelopment, clothing the parts denuded of their serous membrane with a protecting covering against irritation from the air. Just think of it! as if the mucilage were not swallowed down, but remained as if applied with a brush sticking to the larynx or even to the inflamed mucous membrane of the kidneys. What really childish view!—In reality the mucilaginous fluid enters into the channels of the lymph and the blood-vessels and gives to the blood in the lungs the material for the formation of fat and also oxygen in electric tension. The latter accelerates the circulation and removes thereby the inflammation caused by the stagnation of the blood. Gum arabic is besides a very valuable nutriment in the hot season. While it makes the supply of easily decomposed ammoniac food unnecessary by its earthy substance which gives cohesion to the albumen, this aliment not only stills the hunger, but also opposes chemical decomposition, especially as it rouses electric force through its transmutation into fat and the liberation of oxygen. The negroes during the time of gum-harvest live almost exclusively on the gum, and so also while transporting it to the emporiums. Even the monkeys in South-Africa are said instinctively to make use of gum arabic as food, so also the Hottentots and the Bushmen who keep themselves in good health for many days on a daily ration of $1/3$ of a pound of this substance.

Gum arabic is, indeed, altogether insipid, it can not be regarded as a dainty, but this is easily obviated, as we shall see.

But besides gum arabic there are other glucosides of similar efficacy. A very generally used substance is *licorice* which is the inspissated juice of the root of *Glycyrrhiza glabra*, and which contains the glucoside

"Glycyrrhizin" in considerable quantity. In former times licorice-water deservedly played a chief roll among medicines; but the patients themselves are in fault, if this almost most effective of all remedies is no more so generally used by physicians as formerly. The remedy became too well known to the patients and too common, so that they scoffed at it, and now they bear the damage, for the physician now prescribes for them instead of the licorice-water, which supplies vital force, poisonous morphine and water of bitter almonds containing prussic acid. This drives away the irritation to cough so effectively, that the nerves afterwards lie still, as if a board had been laid over the chest and this had been weighed down with stones.

Similar to the effect of licorice-water is the infusion of a mixture of licorice-root and marsh-mallows (Althaea officinalis). Of almost equal value is the vegetable mucilage contained in linseed. All these remedies are known, but the mistake is made in their application that their nutritive value is overlooked, and they are consequently used in too small quantities.

A mixture of equal parts in weight of the roots of marsh-mallows, licorice and linseed for a tea is with good reason favored as a remedy for consumption; only we should remember that *large quantities* of it should be used. Marsh-mallow root contains starch, sugar, vegetable albumen (Pectine, Asparagin) and a mucus corresponding to the glucoside of gum arabic in concentrated form. Almost the same substances exists together in the licorice-root, together with the glucoside Glycyrrhizin with lime and magnesia.

In dwelling upon these glucoside aliments which supply oxygen, I would remind the reader, that these remedies have been known as to their effects empirically for centuries and have been in diligent use. So there is under the name of brown leather-paste a licorice paste consisting of 1 part licorice, 15 parts gum arabic and 9 parts sugar, farther a gum-paste composed of 4 parts sugar, 4 parts gum arabic and 3 parts of the white of eggs. The last is nothing else than an ammoniac glucoside (Hensel, "Das Leben", 2^d Edition p. 35) and is as all glucosides able to give off oxygen. Thus is explained, how the little chick can live and breathe while the egg-shell is yet closed. And is it otherwise with the embryo nourished by the maternal albumen of the blood??—

From this it may be manifest, what an internal source of oxygen we may supply to consumptives by offering them food containing albumen.

Among these aliments rich in albumen are all nuts and most seeds. Especially valuable are almonds, so also hemp-seed and poppy-seed. They all contain the glucoside *Emulsin;* thence it may appear why emulsions of almonds, poppy-seed or hemp-seed were prescribed by physicians with the best results, formerly ten times as much as now. Medicine also is ruled by fashion, and since the blessed régime of the bacilli has come in, the good old effective remedies are more and more forgotten.

Of what value the emulsion of almonds is and, indeed, as a supplier of oxygen by way of the venous blood-channels, may be concluded from much experience: In Carlsbad e. g. they bake an almond-bread for diabetes-patients; for those with diseased lungs they have a kind of gum-paste *in form of powders.* This is made of 8 parts in weight of sweet almonds, 4 parts sugar and 1 part gum arabic.—The licorice-paste in powders was also formerly used, consisting of 2 parts of licorice-powder, 1 part powdered sugar and 3 parts powdered gum arabic.

In Berlin also a glucoside remedy against lung-diseases has been used for more than 100 years, called "Dr. Kurella's Brustpulver" (chest-powder). It consists of 1/2 parts powdered licorice-root, 2 parts senna-leaves, 1 part sulphur, 1 part fennel-powder and 6 parts sugar. This remedy is useful not only for coughs, but also for hemorrhoids and for costiveness. The explanation lies in the nerve-quickening supply of oxygen furnished by the glucosides, and at the same time in the sulphur useful in forming the red blood-corpuscles which absorb oxygen. The glucoside of the senna-leaves is called chrysophan, and fennel-seed contains sugar as well as vegetable albumen; from this it may be believed that the use of "Kurella's Powder" in many Berlin families reaches large quantities; it is bought by the pound, and may really be considered as food rather than as medicine.

That stagnation in the region of the portal vein is best removed through glucosides which furnish oxygen, may also be seen from the bark of the blackberry-bearing alder (*Cortex frangulae*) and the rhubarb-root (*Radix Rhei*) as further examples. Both contain the same glucoside as senna-leaves, i. e. *chrysophan.*

Similar to gum arabic, in the sense of being an aliment which supplies oxygen through the venous system is the gum of tragacanth, and in the same category may be enumerated the mucus from our native salep-roots, the powder of which is of high nutritive value especially with children, and shows its usefulness by forming in the body both oxygen and fat. These excellent suppliers of oxygen for the intestinal nerves in the hot season should be used a 100 fold more than has yet been done, and they ought to be carried in greater quantities on ships conveying passengers. Is there a more lasting nutriment and of greater cheapness than sweet almonds? They present a perfect conserve yielded by nature, for they contain about one half of oily material and one fourth waterless albumen, while the rest consists of sugar and tissue, the latter being combined with lime and magnesia. So the almonds furnish everything necessary to form bones, nerves and muscles, and all that is necessary to make them savory, is to add sugar. Such a mixture has the name of marchpane (lat. *marcipanis*, = *Marci panis*) because the legend declares, and this may be a real fact, that St. Mark for a long time lived solely on almonds which served him as a bread-fruit. With this he could easily remain healthy and strong.

We see then the limits between medicine and nutriment altogether disappear; for the *"Pulvis Amygdalae compositus"* (8 parts almonds, 4 sugar and 1 gum arabic) is only a sort of marchpane in form of a powder. It is similar with *"Pulvis Tragacanthae compositus* (1 part starch, 1 powdered tragacanth, 1 gum arabic and 3 parts sugar).

When mucilaginous substances are mixed with sugar and albumen, we have both aliment and medicine. In this category is also a mixture, of arrow root flour, salep, sugar and cocoa, called "Racahout", as well as an infusion of figs, licorice, marsh-mallows and St. John's bread.

All these nutriments are of use against consumption or phthisis.

To expedite the operation of oxygen in the more advanced stages of consumption, we may use besides the glucosides (gum arabic, licorice, mucilaginous plants) Superoxide of hydrogen (H_2O_2) with appreciable effect. While this preparation supplies the first parts of oxygen, which are necessary to liberate larger quantities of oxygen, it is suitable for all catarrhal affections, but especially in consumption, when used in combination with the physiological effects of "Hensel's Tonicum." The two preparations support one another, the latter by enabling the blood to chemically absorb oxygen, the former by putting oxygen within reach in the circulation itself. For practical use the purchased superoxide of hydrogen, which corresponds to a 10 fold volume of oxygen is mixed with 4 to 5 times the quantity of water; daily twice one porcelain spoon-full is given, or a table-spoonful in a small wine-glass; for when the preparation touches a metal, e. g. a silver spoon, it at once decomposes into water and oxygen.

Although in this manner life-giving oxygen is secured through glucosides and superoxide of hydrogen together with a preparation of iron which forms hemoglobin, and thus the most important requirement is provided for, the other factors which have an influence in normalizing the constitution of the blood must not be neglected. Among these the foremost place belongs to physiological salts, to restore to the serum of the blood its electrizing, antiseptic power. Experience teaches that in all cases of stagnation of blood in the glandular organs (liver, spleen, stomach, intestines, kidneys, bladder, lungs, ovary and uterus) the salt-containing mineral waters of Salzbrunn, Kissingen, Marienbad, Carls-bad, etc. do such good service, that patients always return improved from these places, and that the improvement continues until through a deficient supply of the blood-salts a loosening of the tissues involved and a consequent disturbance in their functions takes place. Such a loosening or spongy consistence of the tissues is to be compared with the state of meat, which is laid in salt, whereby it retains its firm structure, in contrast to meat which is laid in a bowl with clear water, in which its substance becomes loose and is partially dissolved.

If now we examine the chemical constituents of the above-mentioned mineral waters, they contain especially hydrochlorates and sulphates together with iron in varying proportions. It is manifest, that these waters despite their favorable effects do not show those proportions of mineral salts which correspond to the normal serum of the blood; their effects are therefore in all cases surpassed by the effects of the *physiological salt-water*, which day by day assists the serum of the blood in gaining its normal consistence.

This has also its application to those who have diseases of the lungs. What especially benefits them are the sulphates. Formerly when yet under the influence of the teachings of scholastic medicine, I supposed, that properly prepared and completed organic substances must be supplied to the blood and the lymphatic juice, to be assured of prompt assimilation. Now since the secretion of bile and of the pancreatic juice constitute the particular absorbing material by means of which new substances are assimilated from the chyme, and considering that Taurine may be extracted from the gall, and this consists of two carbohydrates (CH_2) with sulphate of ammonia, I formerly used ethyl sulphate of ammonia, (which is isomeric with Taurine) in treating consumption, and, indeed, with very favorable results. Nevertheless, with advancing knowledge I found, that it is only necessary to administer any sulphate, in order to have taurine formed in the liver through chemical transformation. So we may continue to use with equal success the physiological salt-water for normalizing the serum of the blood in consumption.

The hitherto very general application of sulphuric acid in diseases of the lungs, which unconsciously sought the same end, may therefore be dispensed with, and then the disadvantage resulting from the neutralizing of the alkaline constitution of the gall by the sulphuric acid, thus destroying its ability of assimilating fats is done away with.

We have so far learned to know the healing power of the *glucosides* as suppliers of oxygen, *superoxide of hydrogen* as an adjuvant, further, *iron* and *physiological salts* as constituents of the blood which generate magnetism and electricity. We have yet to consider *milk* as the aliment richest in lime, which aids in forming new red blood-corpuscles, which are enabled of themselves to chemically absorb from the surrounding atmosphere the electrizing and nerve-quickening oxygen. It remains to direct our attention to the infectious catarrhal secretions. Although the disinfecting power of the serum of the blood is increasingly augmented, as its contents of salt become normal, it is, nevertheless, useful to take direct measures to stop the putrescent disintegration of the catarrhal secretions of the air-passages. This is effected by the inhalation of vapors of vinegar, or yet better of *formic acid*, in case such can be found in the drug-store. The reasons for this have been given in the pathological part.

Consumptives ought besides to be liberated from the carbonic acid accumulated in the blood, which paralyzes nervous action. We cannot too often remind the reader, that a candle is extinguished if we take it into a room the air of which contains one fourth of carbonic acid. So also the activity of the nerves of the lungs is extinguished, if they are encompassed with blood laden with carbonic acid, and instead of imparting electricity to the connective tissue of the lungs, they leave it to chemical decomposition, or what we call *putrefaction*. Consumptives ought to be sent to the mountains, in order to remove the carbonic acid from their blood; for there they can breathe in cool air, free from dust, and the carbonic acid is more easily exhaled under a diminished pressure of the atmosphere. The least we can demand is, that consumptives should be transferred from the ground-floor to a higher and drier story.

Cramp of the Stomach.—Colic of the Intestines.—Gall-stone colic. The cure consists in applying Hensel's Breathing (see p. 140) and improvement of the constitution of the blood by tonic limonade (see p. 185) and physiological salt-water (see p. 183).

Abortion. The death of the foetus is caused in by far the most numerous cases by the fact that the mother's blood, either in quantity or in quality is not suitable to sufficiently nourish the new organism. Just as pears and apples, chestnuts and fruits of every kind fall off prematurely and undeveloped, if the soil does not contain enough of the substances necessary for the formation of the fruit (potash, soda, lime, magnesia, phosphoric acid, sulphuric acid), so also in the tendency to abortion, the weakness of the constitution of the blood is at fault. This state is successfully combatted by the daily application of a half pint of physiological salt-water, and daily 20 to 25 drops of Hensel's Tonicum on a teaspoon-ful of sugar.

Morphine habit. It is with the use of morphine as with the use of the vapor of stramonium-cigarettes against asthma. The nerves, when once they have lost their sensibility through the narcotic, call for a repetition until the correct physiological means of tranquillization is given them by a normal circulation. Rubbing the body with salt-water, Hensel's Breathing (see p. 140) and daily a glass of tonic limonade and of physiological salt-water, supported by the administration of Calcium-Magnésium-phosphate and flowers of sulphur (see p. 184) gradually remove the morbid craving for narcotics.

Nervousness. God every day makes the sun to rise in the morning and set in the evening. As long as we are exposed to the rays of light, the life of the nerves strives for action, but when the sun has disappeared from the vault of heaven, it desires rest. With this arrangement, which requires a change between rest and activity, we have done well for centuries. But now by means of the electric light we turn night into day, and during the day our nervous system is moved, not only by whatever takes place in our nearest surroundings, but the newspapers take care that the calamities from all cities and countries of the earth occupy our thoughts. Telegrams and telephonic messages prevent our nervous system from taking rest. Distances of time and space are reduced by the telegraph and the railroads. With such ample resources men desire to accomplish correspondingly more than in former times, and even against their will they are drawn into the general whirl of restless activity. The surplusage of influencing factors causes a constant vibration of the nerves. Man cannot withdraw from this, so long as he remains in his respective sphere. I therefore know of no other radical remedy against nervousness but the peace of nature. Out into the country!—Out on the mountains! A palliative is afforded by a dose of 8 grains of flowers of sulphur.

Rheumatism is caused by the checking of the circulation in the capillaries and it requires the same treatment as stones in the bladder (p. 114) namely warm baths to expand the blood-vessels; physiological salt-water to change the urates which are difficult of solution into double salts; sulphur, lime and iron to increase the number of red blood corpuscles.

Sleeplessness. This symptom accompanies qualitative poorness of blood and may be removed as soon as a normal constitution of the blood has been restored. The tonic limonade alone taken in the evening proves an effective sleep-potion in a physiological sense. In Berlin I was once visited by a book-seller whom I had not seen during the 8 years of my absence in foreign parts. I offered him for refreshment a glass of tonic limonade which he drank down as Hungarian wine. As he seemed to like it, I prepared him a second glass. The next day he returned and asked, what it was that he drank with me yesterday. "Did you feel well after it?" I asked, "Yes," was the answer. "Since 3 years this is the first refreshing night's rest that I have had."

It is with the nerves as it is with the whole man. If they receive what they ought to have, they are contented. I had long known the action of tonic limonade in producing refreshing sleep, but I had not mentioned it in my writings, and I was therefore surprized to hear that "Hensel's Tonicum" is used also elsewhere as a rational remedy for strengthening the nerves, and as a physiological sleep-potion. So I received the following letter from Prof. R. in Lausanne, dated March 20th:

Dear Sir, Permit me to have recourse to your kindness for precise indications as to the therapeutic value of the *tonicum* which bears your name, and which has been successfully used by a number of learned men against sleeplessness.—I was sent a bottle of it in which a professor of our pharmaceutical school has found *formiate of iron*. An other chemist thinks that the *tonicum* is like "Bestucheff's Drops."

The mixture is certainly *soporific*, but may it not after a certain time trouble the normal functions of the brain, and produce effects more or less pathological?

One of those who use the Tonicum which he received from Berlin, boasts much not only of its soporific effects, but also its happy effects on morality.

You have doubtless published a pamphlet on this medicine, on its advantages, as also on the disadvantages that may result from a wrong application. Has this soporific been tried on any large scale in the clinics of the hospitals?

Insomnia is a nervous disorder so common at this day, that a sure and harmless remedy would be a real benefit to the cotemporary generation.

Hygiene and *magnetism* with which I have occupied myself for many years, have made me well acquainted with the means for grappling with disordered sleep, whether before or after midnight, but there are some rebellious cases for which it is necessary to have recourse to the pharmacy, taking great precautions with respect to toxic agencies and with respect to the wise combinations of chemistry.

Permit me to offer you my toast to your health! You will see by my publications, that I have long and earnestly occupied myself with the means for combatting the evils which afflict the human family, whether these be physical or moral. To *diminish the pains* of the body and of the soul in the individual and in society ought ever to be the constant occupation of every true friend of progress and of the common wealth. Etc.

It is always encouraging to find that there are fellow-workers who also are intent on the bodily and the moral welfare of humanity. The distrustful glance of the writer at morphine, cocaine and chloral hydrate as toxic agents is well founded.

Sterility. I have in a number of cases observed, that cutaneous eruptions and sterility go hand in hand. The connection of the two is found in the deficiency of sulphur in the blood. If the albumen of the blood does not contain enough sulphur, no vitelline (ovum) containing phosphorus can be separated from it.

The fishes of the sea are of astonishing fertility. If we only consider the inexhaustible multitude of herings and the countless eggs of the roe of fishes (Caviar). The favoring factor lies in the saltiness of sea-water, with which sulphate of lime and sulphate of magnesia are always commingled.

In agreement with this view there are found in certain places "Bubenquellen" (boy-springs), whose water contains sulphates. Weakly women are sent there, to strengthen themselves, so that on their return home, they may conceive and bear progeny.

As a contrast, the inhabitants of the Sandwich Islands are dying out from deficient fertility and—at the same time leprosy is common there (see p. 151).

The number of Sandwich-Islanders was stated by Cook 120 years ago at 400,000. But the census in 1832 gave only 130,000. This number had in 1836 already sunk to 108,000. In 1849 there were only 84,000. In 1853: 73,000; in 1860: 69,000; in 1866: 62,000; in 1872: 56,000; in 1875: 45,000. How many are left now, I do not know. If this extinction is compared with the mortalitiy of other islanders who limit themselves to food poor in sulphur, we cannot doubt, that flowers of sulphur must help against this state, especially since tetanus also points to the sole rule of the nerve substance which contains phosphorus.

In August 1891 the newspapers of Glasgow, Scotland, brought

reports of an epidemy in St. Kilda, one of the smaller western islands; according to these the islanders lose most of their children within a week after their birth, of tetanus. The first born child is said to be usually exempt from this; which may be explained by the fact that the mother's blood uses up all the disposable sulphur-containing albumen for the formation of the first child. One of the families has lost 12 and others from 2 to 8 in this manner. This mortality was ascribed with reason to their manner of living, and indeed to the oily nature of their food which consists chiefly of birds and fishes. Rev. Fiddes, who sent the report remarks, that no remedy has as yet been discovered, though the peculiar malady has been reported to the medical men of Glasgow."—On my part I would recommend besides sulphur also iron. Such a treatment in a case where after 20 years' marriage all hope had been given up, was followed by the birth of a daughter, causing great joy.

Dropsy. The alkaline reaction, due to Ammonia, of the fluid accumulating in all kinds of dropsy and which in other respects is like the serum of the blood, gives the clue to the process which lies at the bottom of dropsy. Whether great losses of blood, chills and fever, customary indulgence in alcoholic drinks, defective nutrition or chronic diseases precede dropsy, a faulty constitution of the blood is always the deciding cause, and especially the deficiency of salts in the blood is to blame. My theory is the following:

As all secretions take place under the direction of the nerves, so also that of the kidneys. Now as to the plexus of the kidney-nerves, which mostly spring from the sympathetic nerve, we know, indeed, that its substance contains both sensitive and motory elements, but the sensation is transferred only slowly and fades away as slowly. This may be made visible as to its principle by a mechanical irritation of the intestines, say in that of a rabbit just killed. If we scrape across it with the thumb-nail, we at first notice nothing at all; only after a few seconds the wall of the intestine contracts at the place touched, in a comparatively deep furrow, and this furrow remains visible for some time; but gradually it disappears again, and the intestine regains its former appearance. All the intestines have spiral nerve-fibers of the sympathetic nerve running through their walls, somewhat like the wire wound round a metallic string of a violin; the walls of the blood vessels are constructed in the same way, so that the property of the bloodvessels of expanding and contracting is due to the slowly sensating and slowly moving nervous fibers, which as shown in the anatomic part, have their origin in the spinal marrow. This slowness in the function of the sympathetic nerve is owing to the fact, that it does not spring from its own electric quality, but from an indirect source, and indeed from the electric current which flows through the adjacent walls of the blood-vessels, which excites in the sympathetic nerve branches a secondary or inductive current. The electric fluid for the blood-vessels is generated immediately through the salty contents of the serum of the blood, and it may be seen at once, that when these salty contents are diminished, also the electric current in the walls of the blood-vessels diminishes, and corresponding therewith the intensity of the sympathetic nerve-function. This diminished electricity first manifests itself in the *kidneys*, because this glandular organ has the mission to excrete from the blood the urea formed by the oxidation of the tissues, which on the other hand has the property of entering into a chemical combination with an equivalent of common salt or of a sulphate or a phosphate. This explains why the urinary secretions show double the quantity of salt as the blood. Men not chemists have thence falsely concluded, that the kidneys have the function of excreting the useless salts from the blood, while yet the salts in the urine have only the purpose of forming double combinations with the urea which preserve it and make it *harmless*. The natural conclusion from this is, that the salts of the blood must be replaced in the same proportion as they in company of the urea leave the body, so that the urea may find enough salts with which it may enter into harmless double combinations. In keeping with this, the proverb says: "The salt spoils the victuals, if it is not put in." Every shepherd also knows that his sheep get sick, if they have no salt to lick. The reason of this is, that also the urine of the *herbivora* contains urea, which requires salt, in order to form harmless double combinations; else there arise epidemies among the cattle.

Now if the contents of salts in the blood, which flows to the kidneys is diminished, e. g. through drinks poor in salt like wine or beer which dilute the blood, then the injury results without fail, that the electric fluid in the capillaries of the renal arteries, at first imperceptibly, gradually more and more perceptibly suffers a correspondent weakening, and at the same time the attractive force of the ramifications of the nerve-plexus of the kidneys with respect to arterial blood and its repulsive force with respect to the venous renal blood is diminished. This occurs with the more certainty, if with insufficient contents of salts in the blood, the urea arriving in the kidneys with it, from its decided inclination for forming double salts, deprives the nerve-substance of the salt proper to it (ammonium-phosphate) see p. 17 and leaves the remainder as a passive fatty substance (fatty degeneration of the kidneys)

When the latter stage has come, all help is too late; the dropsy which sets in in such a case is always fatal, because with the stoppage of the function of the kidneys the urea which remains in the blood becomes carbonate of ammonia, which lames the nerves and decomposes the blood; especially also the iron of the hemoglobin is separated as an inorganic oxide, and thus the capacity of the blood for absorbing the life-giving oxygen is destroyed.

The symptom of drawing water from the venous blood is also to be ascribed to such liberated carbonate of ammonia, as also the contents of albumen in the dropsical fluid is explained by the capacity of the carbonate of ammonia of dissolving albumen.

The fact that the *feet* first show the symptoms of dropsy, is owing to the above-mentioned peculiarity of the nervous fibers of the sympathetic nerve, of conducting sensation and motion only slowly. It is manifest that under such circumstances the contents of the capillary blood-vessels which are most distant from the muscle of the heart (which in great part is woven from sympathetic nervous fibers), first fall as victims to this separation.

In this first stage of dropsy for centuries acetate of potash (thus an electrizing salt) in connection with squills *bulbus scillae*, (which present a real magazine of gum-like mucilage, thus of glucoside, which according to page 159 is able to supply oxygen*) has brought aid. Nevertheless the use of physiological salt-water, interchanged with sweetened mucilage of gum and tonic lemonade seems more natural and according to my experience proves best. By these means 1. the urea receives mineral salts for the formation of double salts, 2. by the formic acid in the tonicum the paralyzing ammonia is neutralized, while at the same time new iron can become operative in organic form, and 3. by the glucoside, gum arabic, an internal source of supply for oxygen is presented. By allopaths, besides *Kali aceticum*, also the mixture of double salts from borax and tartar, called *Tartarus boraxatus* is used with decided effect. The effective principles are especially the free acids of the bi-tartrate of potassa and the bi-borate of Soda which neutralize the injurious ammonia. To this is to be added the electric chain originating from the common presence of potassa and soda salts, and the well-known antiseptic influence of the borate which prevents chemical decomposition. In connection with a preparation of iron the use of *Tartarus boraxatus* for dropsy, the bacillus of which has so far as yet been hidden from the cognizance of bacteriologists, must be judged to be thoroughly appropriate.

Also lemonade acts favorably, supporting the cure, but still better is sugar-water with cream of tartar. Both drinks neutralize the injurious ammonia, and the bi-tartrate of potassa adds beside the electrizing operation of the salts. The latter effect ought to be sought for in dropsy also externally by daily rubbing the body with salt-water 2 grammes, (31 grains) of common salt to 250 grammes (1 pint) of water.

Finally we would recommend in dropsy the use of fat, because

---

*) This supply of oxygen is so intensive, that smaller animals which of themselves have a more rapid circulation, as mice and rats, after eating squills die of palpitation of the heart, analogous to the fatal effect of chlorate of potassa on children.

also the carbo-hydrates combine chemically with urea. There is e. g.
Ethyl-urea, Di-ethyl-urea, Tri-ethyl-urea and Tetra-ethyl-urea, and all
these combinations unite together just as urea alone combines with one
or more equivalents into salts. The use of sweet-oil, every two hours
a tea-spoonful, with some acidulous beverage (lemonade or sugar-water
with cream of tartar) rests on a safe physiologico-chemical basis. Thus
the further appearance of free urea in the blood may be prevented,
and even the carbonate of ammonia which is generated from urea by
its taking up one equivalent of water, may by combining with oil become new gelatine, (see p. 82). This also explains why dog's fat, which
is used by the common people against consumption (with the idea that
it is so effective because dogs are never consumptive, while this merely
rises from their continual trotting about and breathing) shows decidedly
beneficial effects, even if the druggist should hand out *Axungia porci*
instead of *Axungia Canis*; since all natural fats are based on a 7 fold
group of hexylenes (C_6H_{12}) and the essential difference consists merely
in this, that the one kind is combined with 8, the second with 10 and
the third with 12 atoms of oxygen.

With respect to the *origin* of dropsy, I would like for illustration
to adduce a few facts from experience. Although single examples may
not be allowed to lay claim on general demonstration, nevertheless, the
physiological and pathological process in their principles may be therein
studied. Something then out of my practice:

A man of somewhat advanced age drinks during the day much
milk, in the evening *weissbier* (a thin beer). The consequence is a
bloated abdomen, costiveness, oppression of the stomach, loss of appetite,
heavy feet, pains in the calves, rheumatism, dizziness, headache and
sleeplessness. The flesh of the calves becomes soft and doughy; he discharges but a little thick reddish urine. This would constitute apparently 10 symptoms, but in truth only one: lack of salts and iron in the
blood, whence the urea accumulates in the blood.

Cows' milk contains but 93 to 108 grains of physiological salts to
the quart (liter), while blood requires 123 grains. Light beer (weiss-
bier) has merely 8 grains per quart, thus only the 16th part of what
is required for the blood. Now since with every quart of urine 278
grains of salts are secreted, the contents of salts in the blood steadily
retrogrades, if, as is usual with those who drink milk, little other food
is partaken of. How long will it be, before the 62 grains of iron and
the 1852 grains of lime, potash, soda, magnesia and manganese, which
we have in our blood are exhausted with such a diet? Then the spleen,
kidneys, liver, stomach and intestines are relaxed on account of the
watering of the blood, and all secretions cease. The lack of iron causes
a venous characteristic i. e. one poor in oxygen, with the result of *stagnation of the blood,* which is felt in the different parts of the body
partly as twitches (rheumatism), partly as oppression, partly as a dull

pain. In this case the administration of a preparation of iron easily assimilated (every two hours 20 drops of Hensel's Tonicum, dropped on a tea-spoonful of granulated sugar), and of physiological salt-water (1 pint a day) is indicated. But what was done by our scholastic medicine? *They fought against the symptoms!*—The pains in the head and in the thigh originating in a stagnating circulation were treated with *morphine* which paralyzes the nerves!—Sleeplessness was treated besides with *chloral*-hydrate!—Costiveness and the accumulation of dropsical fluid were met with drastic *jalap*, which causes, indeed, watery stools but cannot remove the cause: deficiency of salts and of red corpuscles in the blood.

How much excess in beer weakens, we see in the cab-drivers of Berlin. All of them know, where good *weiss-bier* (thin beer) is on tap, and the greater number of them is so bloated and weak, that they can only handle the trunks of travelers with difficulty. The Bavarians and Suabians indeed also drink a good quantity of beer, and still retain their health, the reason of this is: they are accustomed to eat radishes and salt with their beer. Since radishes are very rich in sulphur, lime and iron, these substances which together with common salt are lacking in the beer are supplied to the blood in the radishes and salt.

As with the beer, so with the wine. A patient told me, that he had been in Luxemburg, where the wine is good and cheap. So he thought, he would indulge himself in a good treat; but after 3 months he was not able from weakness to stand on his feet, which as is well known is also a specific symptom with drunken people. An other instructive example is the following: A few weeks ago I was traveling in the same railroad-coupé with a young man who introduced himself to me as the traveling agent of a wine-house. Jokingly I told him: „More people drown in wine than in water." I, indeed, also have wine in my cellar, which patients often send to me, but I only drink it when friends visit me; and to pass away the time, I gave him the physiological reasons therefore. He listened attentively and then confessed to me, that his brother had lately died, who had owned an independent wine-business, and who, to animate his guests, found it necessary to drink a good deal of wine himself. He however eventually fell into an incredible weakness, and his death ensued from dropsy, cercocele and apoplexy. His blood was quite black. I told him, that this exactly confirmed, what I had said: the blood had been in the end completely lacking in salty constituents, in sulphur, lime and iron. It had therefore lost the capacity for chemically absorbing oxygen which quickens the nerves, and for rendering the urea harmless, so that the blood received the venous character, laden with carbonic acid and distinguished by its dark color. The gathered carbonic acid in the end burst the walls of the cutaneous veins of the lower thigh, which had become relaxed and loose through the lack of salt, and after a channel had thus been opened, the entire venous blood, pressed upon by the accumulated carbonic acid swept out in an uncontrollable current. The consequent lack of blood in the brain had to end in apoplexy.

As to the curative treatment in such cases, it must be considered that by an excessive use of wine gradually not only all the sulphates and chlorides of the blood are washed out with the secretion of the urine, but the albumen of the blood, which is essentially an albumen resting on soda and lime also suffers a degeneration. For the ashy constituents of wines constitute in the most favorable case only $4/10$ % thus not even a half percent, and this small quantity of ashes consists as to 80 or 85 % of potassa and Phosphoric Acid. Thus in wine not only the sulphates and chlorides are lacking, which are neccessary for the normal serum of the blood, but also the earthy constituents which are necessary for the new formation of the hemoglobin consumed through respiration. The latter as the vehicle of the oxygen breathed is indispensible in every kind of nervous function, and it must be systematically augmented through the presentation of *physiological earths* while the serum of the blood must be brought back to its normal constitution through *physiological salt-water*.

Physiological Earths!—Such are the combinations of fluoric, phosphoric, sulphuric and silicic acid with lime magnesia, and the oxides of iron and manganese in a suitable proportion.

The more subtilely divided these physiological earths are presented for the purposes of digestion and assimilation, the more easily can they be assimilated through the gelatine-sugar (glycocoll) of the blood-gelatin into organic substances, and the more reliable will be their curative action. In this respect we wish to acknowledge the conscientious preparation which distinguishes the trituration of these physiological earths with the four fold amount of sugar of milk as furnished by the firm of Boericke & Tafel in Philadelphia in samples submitted to me. The use of these earths accompanied by the simultaneous use of physiological salt-water will always be effective, not only in *dropsy*, but also in *anaemia, chlorosis, catarrhal affections, spasmodic complaints, swellings of the lymphatic glands, jaundice, yellow-fever, intermittent fever, falling out of hair, excessive obesity and rheumatism.* Not too much is expected, when we assume that by means of the physiological earths and of physiological salt-water the use of any single substances such as preparations of iron and lime for the improvement of the blood in all affections of the nature of dyscrasy will be in future unnecessary, since the former offer *everything* that is necessary to restore the blood to its normal constitution.

With this I shall conclude the therapeutic part. Since this book is specially intended for practical physicians, I have limited myself to the treatment of diseased states, which are as yet involved in darkness owing to the lack of knowledge respecting the chemical changes involved, and since my simple method of cure, which amounts to normalizing the *constitution* and the *circulation* of *the blood* has shown the most manifest practical success, I have thought, that I ought to put it at the disposal of the younger medical generation. It is not only suitable to the diseases specially enumerated, but in general to all kinds of diseased states, and is also suited to veterinary practice.— Would that it might put an end to the widely spread, now almost general, yet really childish doctrine of Bacilli and Bacteria as causes of maladies!

We may indeed reach Rome by various ways, but by the search after bacteria the healing art is not advanced one hair's breadth. This can sooner be done by massage and Hydrotherapeutics, for by massaging the stagnation of the blood is removed, and hydro-therapeutics acts in two directions, i. e. first, electrizing through difference in temperature and secondly, through chemical modification which dehydrated organic substances that thereby have become fixed like fibrin, experience through the re-absorbtion of water. In a gross way this might be compared to the manner in which the ends of the toe-nails which have become hard and at last shiver like glass become yielding and flexible through a warm foot-bath; just in. the same way it is with cartilaginous and tendinous substances. But this in no way exhausts the essence of hydro-therapeutics: we must also yet consider with respect to the connective tissue and the muscles the chemical changes, which result from the detachment of the elements of water (HHO) on the edges of the molecules of creatine, sarkine etc. (see p. 67); by this a state of torpor is caused which may be removed through a chemical re-attachment of water. But, nevertheless, sulphur, lime and iron and the blood-salts cannot be thereby supplied to the body, if they are deficient.

As to other methods of healing, I have been advised from various quarters that Schuessler's therapy has the same basis with mine; but I must state that Schuessler's therapy only superficially seems to resemble mine by a faint glimmer, in this that it also uses the mineral substances found in the body, but in the greatest dilution, while I offer the same in sufficient quantities as nutriments for the blood-globules and the serum. It is self-evident, that if the bones of a patient have lost in the course of a year 3000 grains of phosphate of lime and of magnesia through the blood-vessels passing through them, that this quantity cannot be restored by administering the one trillionth part of a grain. What does the reply amount to when it is claimed: "The rest is furnished through nutrition, for which there is a renewed appetite;" if the aliments are such that they contain no lime, as is the case with meat, wine and beer?

As to the rest, Schuessler's "Abridged Therapy" is built up on wholly obscure and in great part false hypotheses. Especially is Schuessler's idea that the individual diseased cell must be restored to health, utterly untenable. The individual cells must in no way be preserved, but on the contrary they must be subject to a continual consumption and mutation of substance. Remaining in good health rests on the *renewal* and *transmutation* of the whole organism. Unless substances are used up, there can be no renewed growth, and the peculiarity and law of animal life lies just in this, as indeed Schuessler, contradicting himself, concedes, when he says: "By the side of the origin of *new* cells, the destruction of the *old* ones by the influence of oxygen takes place." Nevertheless he rests on the declaration of Virchow: "The essence of disease is the changed cell."—To this he adds the following:

"Every normal cell possesses the ability, to absorb substances and to reject them. This capacity is diminished or destroyed when a cell in consequence of an irritation has suffered a deficiency in one of its salts. The *status quo ante* is restored, when the deficit has been covered by homogeneous material from its immediate soil of nutrition. [Concerning this "soil of nutrition" Mr. Schuessler gives no information.] When the *changed* cells through the restoration of what was lost have again attained their integrity, they can again functionate normally."

From this statement it appears manifest, that Mr. Schuessler seeks to preserve the *diseased* cell, when yet nothing more salutary to it can take place, than the chemical death by fire, through the oxidizing respiration while its substance is transmuted into carbonic acid, water, nitrogen and urea.

How imperfectly Mr. Schuessler is informed as to important chemical fundamentals, may be seen already from the first page of the introduction to his work, where he repeats the sentence from *Moleschott*, who is not without merit as a pioneer; but this sentence must have been caused by the defective construction of a sentence, namely: "Without a foundation yielding gelatine no real bone can be formed;" in the spirit of Moleschott this should be: "Without a foundation yielding gelatine and phosphate of lime no real bone can be formed."

This only incidentally. When Schuessler furthermore in reply to Dr. Ring of Ward's Island, New York, says, that the cartilaginous tissue is allied to the mucous tissue, this is not plainly false; but when he ironically adds: "Will any one attempt to heal a cold in the head, a disease of the mucous tissue curable by chloride of sodium, with prepared cartilage?" I have to answer, Why not?—The cold of the intestines (intestinal catarrh) is actually healed by cartilage-broth from calf's feet; the idea, therefore, of curing a diseased tissue with the substance of a kindred tissue is not at all as queer as Dr. Schuessler supposes. That organic substances are to be excluded from our list of remedies, as he dogmatically sets up, we find disproved by the effects of citric acid, gum arabic, licorice, the starchy mucilage from potatoes, rice-gruel, oat-meal gruel etc.—Quite peculiar is his statement, that every mineral substance has special doors of communication in the walls of the capillary vessels. How will he make us believe that?— Has he observed it with the microscope? Mr. Schuessler makes numerous affirmations which cannot be controlled as to their correctness through any physical and chemical facts. He says: this is so, but he brings no proof and asks for a blind faith. He says, e. g.: *Concussion of the brain.*—The depression of the function of the respective braincell will require *Kali phosphoricum.*"—With these two lines the subject of "Concussion of the brain" is put off. This is indeed a *very* much abridged therapy, and yet it is not abridged enough; all the unprovable and unproved assertions should first be struck out.

Demonstrably erroneous is the statement that in healthy men, animals and plants the mineral substances are contained in dilutions which about correspond to the third, fourth and fifth medicinal decimal dilutions, for the body of an adult of 150 pounds contains 6 pounds of mineral substances, one quart (liter) of blood contains 123 grains of mineral substance. The grains of cereals yield between 1.6 and 2.2 of ashy constituents from 100 in weight; sugar-beets yield 1.05, horseradish 2.5 per cent.

Erroneous also is Schuessler's belief that *silicic acid* is *insoluble;* it is dissolved, according to the temperature in 200 to 1200 parts of water, and consequently also in warm blood.

If such assertions, which he could easily have corrected from chemical text-books are erroneous, what he says of the processes in the tissues which neither he himself nor another can see or control, deserves no belief.

To the objection that we consume enough *fluoride* of *calcium* in our milk, enough *common salt* in our soups, enough *phosphate of potassa* in meat and in all vegetables, he deems it sufficient to answer:

"One might think, that the molecules of the salts administered as medicine would join the homogeneous salts contained in the blood, and thereby make the cure illusory; but this apprehended union is not effected, because the carbonic acid present in the blood serves the molecules of the salts as an isolating medium."

To this we can only answer: "A good excuse is worth a penny." Everything going into the stomach and into the intestines, so far as it is made use of in the re-building of the bodily substance is absorbed by the chyle-vessels, i. e. the lymphatics, and passing through various lymphatic plexuses it is transformed for the uses of the preservation of the body, and forwarded through the thoracic duct into the venous blood, when it enters the right half of the heart in which a thorough mixing process takes place, to which the valves of the heart largely contribute, by forcing the current of blood to divide. Now suppose we allow the medicinal molecules of Schuessler to have advanced thus far. Does he overlook the fact, that the carbonic acid, which should isolate and thus protect his molecules, is dissolved in gaseous form in the blood, and not present in prickling bubbles, and is besides mostly excreted in the lungs? Does he overlook furthermore, that the serum of the blood which dissolves the salty particles has no elements formed like "cells" at all? To what saving shore of cells could then the salt-molecules of Schuessler, isolated by carbonic acid commingled in the general serum of the blood, work their way through the hissing breakers in the heart, and how does Schuessler conceive that these salt-molecules which go first into the lymph, then into the right heart, then into the lungs, then into the left heart, then through the aorta and lastly through its thousand ramifications in the capillary vessels, —how can they manage, so that they may, separated from the homogeneous salty particles of the blood, find the right hole intended for them by traversing the pores of the capillaries?

"All honor to your word, Sir John, but I must have a better surety."

Schuessler has published his "Abridged Therapy" already in a 16th edition. Can this alone be a sign of its excellence, so long as there is nothing else that looks more plausible?—Schuessler is already vanquished by the book of Pfarrer Kneipp, whose "Water-cure" has already seen (1891) the 32d edition.

I like Pfarrer Kneipp's method in comparison with Schuessler's just in proportion to their editions, namely twice as well.

A pity, that Pfarrer Kneipp does not possess chemical knowledge, he might then do even more good, and in his reports he might sometimes better appreciate their significance. He tells us e. g. in his 32d edition p. 64 of a farmer who was at first benefitted by drinking salt-water, but afterwards he was afflicted with stomach troubles, indigestion and anaemia and died of exhaustion. The conclusion to be drawn from this seems very manifest, namely that men cannot live on common salt alone, and that also other substances are needed, to give to blood a good constitution. Instead of this Pfarrer Kneipp falls into the error of conceiving that mineral waters are something improper and sharp, and compares their application to the attempt of scouring golden vessels with sand; at first they would become bright, but soon they would be scratched. He holds, that the vessel of our stomach is much nobler than golden or silver bowls and must be treated softly and gently. This last conclusion is altogether praiseworthy, although we are in no way safe (since everything rises and sets) but that another priest may rise and re-introduce some nostrum esteemed in the Dark Ages. Then the monks said: everything that is used must now and then be purified, so also the stomach. And they invented brushes for the stomach on a long handle which they thrust through the gullet into the stomach, and by turning the brush round and round they instituted a scouring process for the stomach. All Europe imitated the monks. Although we now know that our internal and external membranes cleanse and renew themselves, and that the stomach especially throws off its envelop of mucous membrane in digestion, and during its rest renews it by means of its lymphatics, no priest is in duty bound to know of this physiological process, and any priest of the future may therefore boldly fall back again on the stomach-brush.

So we cannot require of Pfarrer Kneipp that he should know of the chemical and physiological significance of licorice and marsh-mallow, the use of which he (p. 112 in 32d ed.) calls in doubt. Just as little does he need to know that the *mineral substances* he is afraid of, are contained in the form of potash, soda, lime and magnesia salts in all the herbs, even in the oat straw, the hay-blossoms and the pine-leaves, which he uses in his baths, so that he without knowing it, is using mineral baths. Perhaps he has even overlooked the fact that the first member of his medicine-chest, *alum*, is truly and properly a mineral. But all this does not matter. His method operates by this, that through differences in temperature he, unconsciously calls into operation electrically exciting factors, and that he opens the way for circulation. He also (again unconsciously) provides the body through charcoal, chalk and phosphate of lime (bone-dust) and also through bran-bread with the mineral substances necessary for the blood; he also supplies through honey oxidizable, warming substances for the blood, which for obese individuals, who are afflicted with catarrhal diseases is quite rational. And so his method of cure, which strives for the same objects as mine, is only to be praised, although many things that can be secured more easily and more simply are procured by him in quite a roundabout way, and although for certain diseases e. g. inflammation and consumption he furnishes a parable but no physiological explanation.

In inflammatory conditions Pfarrer Kneipp uses the infusion of herbs which without his intending it, act not only by their nerve-quickening ethereal oils, but also through their *tannin* which checks fermentation, e. g. wormwood and sage. In Epilepsy he makes an external application of electrizing salt-water. This is quite correct and it corresponds with one half of his praiseworthy program, that with epileptics the treatment looks to *improvement* of the blood and regulation of the circulation; but Pfarrer Kneipp does not tell us how the blood may be improved.

Such little shortcomings excepted, Pfarrer Kneipp's curative method, so far as it concerns his own independent observations and not the additions which in order to make up a book he has collected from various herb-manuals, may be adjudged to be one of the ways that

12*

may lead to Rome. I say "may", for patients love to meander into
manifold by-ways, which lead off from the goal, and for this the method
is not then to be blamed.

I did not wish to avoid answering the many questions asked, how
I regard the therapies of Schuessler and of Kneipp, but have now
made known my views by this critical discussion.

LIST OF THE CAUSES OF DISEASES.

1. Insufficient Nutrition.
2. Insufficient or too warm Clothing.
3. Damp dwellings, lack of sunlight, corrupted air (loaded with carbonic acid) or an air too hot and filled with dust in the work-room or
the bed-room.
4. The Passions.
5. Atmospheric Influences.
6. Want of bodily or mental Activity.
7. Overwork or lack of sleep.
8. Excessive Eating, as well as' abuse of alcoholic liquors (wine,
beer, grog, whiskey).
9. Smoking of Tobacco.
10. Sexual Excesses.
11. Morphine and Cocaïne.
12. Infection (Inoculations à la Jenner, Koch, Pasteur.)—Syphilis.

LIST OF THE MOST EFFECTIVE PHYSIOLOGICAL AND HYGIENIC REMEDIES.

Almond-Confects. Four ounces peeled sweet almonds, 2 oz. sugar,
1/2 oz. gum arabic and 1/4 oz. gum tragacanth are finely crushed in a
mortar; this serves for making almond-milk.

Almond-Milk. One ounce of almond-confect is rubbed with a little
water to a thin consistency, half a pint of water is then stirred in and
the whole is passed through a piece of muslin.—This is used in fever,
inflammations, consumption; interchanging with gum-lemonade and physiological salt-water as beverages.

Bitter essences. Also the bitter essences like tannin belong to
the glucosides, i. e. they are substances, which have arisen by poly-

merization (chemical multiplication) from glucose, through a chemical splitting off of water after the analogy of the bitter caramel produced from sugar. Bitter essences, tannin and gums—these three substances are not unfrequently together in plants, not infrequently united with bitter resin, representing in this union 5 different stages of oxidation of polymeric sugar. Examples of this are myrrh and aloes which contain at the same time tannin, bitter essence, resin and a still considerable portion of unoxidized gum. Corresponding with the origin of bitter essences from aggregated sugar molecules they also possess the property of the glucosides of changing into fat while liberating oxygen. This peculiarity explains their physiological action. As they give off oxygen in electric tension, this is communicated to the tips of the sympathetic nerve touched by them and even to those of the cerebral nerves, for the effect is already perceptible on the tongue which receives its nerves from a branch of the trifacial nerve. The evidently electrical effect which appears accompanied by a shudder and contorsion of the face from intensely bitter substances e. g. without exception from coloquints, puts the electrising property of bitter essences out of doubts. Owing to this effect upon the sympathetic nerve fibers, they enable these to draw in arterial blood by means of the capillaries, just as a stick of sealing-wax made electric by rubbing attracts light movable objects. From this appear the efficacy of bitter essences on all the glands interpenetrated by the sympathetic nerve, and which perform their functions imperfectly owing to the checking of their circulation in the capillaries. The use of bitter essences extends, therefore, to the liver, spleen, kidneys and uterus, as well as to the intestines, the stomach and the lungs. Agreeably to this, the bitter wormwood (absinthium) proves a general quickening remedy for the nerves, and its influence extends even to the spleen, especially as besides the bitter essence of the herb, also an ethereal oil takes part in this quickening of the nerves. Wormwood cures fever, like the bitter essence of the willow-bark and cinchona-bark, by restoring the functions of the spleen and the kidneys.

A similar effect is produced in liver troubles by the leaves of the marsh-trefoil (trifolium), which even shows some successes in diabetes.

That gentian animates the stomach to renewed activity is known to the common people.

Belonging into the same family with gentian is the lesser centaury (Centaurea) which has a similar action against the inertness of the uterus as Roman chamomile (anthemis). Both are used for the restoration of the menses when suppressed by cold or emotions. The bitter aloes and myrrh serve a similar purpose. An infusion of the bitter essence and the tannin of orange peels, wormwood, marsh-trefoil, gentian and cascarilla root possesses a well-founded reputation as the Compounded Orange-elixir in the deficient function of the digestive organs, and even a few drops have the desired effect, but always only transiently, as without the new formation of blood corpuscles and the renewal of the serum, the intestines cannot perform their functions normally.

Lastly among the bitter essences we have to mention quassia, from which the glucoside quassine $C_{10}H_{12}O_3$ may be formed, corresponding to the glucoside aloïn, $C_{15}H_{16}O_7$ from aloes. Even very weak cold infusions of a little quassia-wood steeping it only 2 minutes, have surprizing effects in giving tension to the nerves of the intestines showing increased circulation. By this action of the quassia it is manifestly proved, that it needs but to touch electrically a limited portion of the plexus of the sympathetic nerve, to cause an electric excitation of the whole system, but in agreement with the peculiarity of the sympatheticus the extension of the excitation requires a somewhat longer time than is the case in the domain of the cerebro-spinal nerves. We must however note the experience for quassia as for "Nux vomica" which emulates quassia in bitterness and efficacy, that the action ceases whenever the oxidation of the bitter essence, changing it into carbonic acid and water, is completed. The most natural and lasting electric quickening of our collective nervous system can only be effected by the normal salts of the blood and the oxygen conveyed by the red corpuscles of the blood.

Calcium-Magnesium-Phosphate. In connection with the use of Hematite and flowers of sulphur together with physiological salt-water in asthma, anaemia, chlorosis, epilepsy, scrofula, eruptions, glandular swellings, catarrhs, obesity, nervosity, gout, rheumatism, consumption, dropsy and general dyscrasies.—Take daily twice, stirring into a cup of salt milk or coffee, or into a plate of any soup that may be eaten, 15 grains.

Citric Acid, $C_6H_8O_7$—With respect to the products into which it may be chemically divided; citric acid is to be viewed as triply condensed glycolic acid (COO, CHH, HHO) and thus as compounded of COO, CHH; COO, CHH; COO, CHH, HHO. As a tribasic acid it can combine with 3 equivalents of Ammonia or 3 of urea. It is to be preferred in the sick-room to fruits which contain malic acid, which acts as double glycolic acid (COO, CHH, CHH, COO, HHO) and acts bibasically, because the latter in the digestive canal easily decomposes into carbonic acid and lactic acid, while citric acid forms with Ammonia 3 groups of glycocoll, which may be used in forming albumen of the blood for new bases of flesh. The most useful way of using citric acid is in the form of gum-lemonade. *Lemon's better*

Cream of Tartar. In the juice of the grapes bi-tartrate of potassa (cream of tartar or tartar) is found combined with sugar and in soluble form; by itself it is but little soluble in water and still less in the presence of alcohol; in proportion, therefore, as the sugar of the new wine ferments into alcohol and carbonic acid it is precipitated from the wine; but in sugar-water it again becomes soluble, for the turbid milky mixture of 15 grains of cream of tartar and a pint of water becomes clear when $^2/_3$ of an ounce of sugar is added and stirred for a little while.—By stirring cream of tartar into sugar-water and allowing the undissolved tartar to settle, we therefore receive a lemonade containing a sufficiency of cream of tartar. More sugar-water may afterwards be added to the tartar settled in the glass.—We thus secure an anti-septic beverage combining with ammonia and urea; useful in fever and in dropsy.

Formic Acid. Anhydrous formic acid $CHHOO$, is crystallizable, like acetic acid. The officinal formic acid (spec. grav. 1.060) contains 25 percent of the anhydrous substance. In this degree of concentration it may like the officinal muriatic acid, which also contains 25 per cent of the anhydrous acid be used mixed with sugar-water in all states of feverishness. We give 10 to 15 drops of the formic acid with one tumbler full of sugar-water. Its character of a volatile destillizable fluid recommends it for inhalation in consumption and in catarrhal affections, in order to neutralize the ammoniacal products of decomposition. Used internally it enters into chemical combination with the urea circulating in the blood and can ward off a transmutation into carbonate of ammonia.

Gum-lemonade. Three heaped tea-spoons of gum-lemonade powder are dissolved in a glass of water.—In consumption, dropsy, decomposition of the blood and inflammation it is to be used alternating with physiological salt-water. Among the diseases showing disintegration of the blood we must reckon diphtheria.

The powder for gum-lemonade is made of 40 parts sugar, 6 parts of prepared gum arabic and 1 part citric acid. Ships which sail in the regions afflicted with yellow fever should have a sufficient supply.

Hematite. After hemorrhages, abortions, in painful menstruation and in troubles in the climacteric period, so also for hemorrhoids, in chills and fever, in rheumatism and gout, as well as in cases of paralysis, in asthma and cramps, in diabetes and general dyscrasy, daily once 4 to 8 grains of hematite stirred into a spoonful of water with a pinch of common salt or in a wafer or a gelatine-capsule; it is best taken before dinner.—In all cases 8 grains of flowers of sulphur and daily half a pint of physiological salt-water should be used.

Hensel's Breathing see p. 140.

Hensel's Tonicum. The preparation is as follows: equal equivalents of the sulphate of the protoxide of iron and sulphate of the oxide of iron in a common solution are decomposed by 4 equivalents of dissolved formiate of lime, using so much water that 1000 grains of the solution contain 15½ grains of metallic iron. To this we add 1235 grains of acetic acid, 7630 grains of alcohol and 154 grains of acetic ether. After the sulphate of lime has been precipitated, the tonic is filtered. The tonicum is used for making the tonic lemonade.

Licorice-drink. Powdered licorice, gum arabic and sugar, of each 231 grains, together with 62 grains of physiological salts are dissolved in a pint (½ liter) of water. It is a healing beverage in catarrhal affections, in cough and consumption.

Lime, Precipitated Carbonate of Lime. In cholera-morbus of children give 8 grains stirred into a spoonful of sugar-water. To be repeated whenever the diarrhoea returns. In dysentery of adults give daily three times, 31 grains each time.

Nerve-salts $= 3(P_2O_5N_2H_6)$. To be used in addition to Hematite in disturbances of the circulation and the troubles resulting thence, e. g. in headache, costiveness, hemorrhoids, bloody flux, rheumatism, diseases of the liver and the kidneys, neurasthenia, tooth-ache and diabetes.— Daily once or twice 15 grains dissolved in a tumbler-full of water.

Physiological Earths. These contain amorphous silica, calcium-magnesium-phosphate, iron, fluoride of calcium and sulphur. To be taken once or twice a day, 15 grains stirred into a cup of salt milk. To be used in addition to physiological salt-water in eruptions, anaemia, chlorosis, catarrhal affections, obesity, falling out of the hair, cramps, paralysis, consumption, swelling of lymphatic glands, gout, rheumatism and diabetes.

Physiological Salts. These consist of phosphates, sulphates, hydrochlorides and carbonates of potassa and soda corresponding with the analysis of the blood-salts made by Denis. They serve for making the licorice-drink and the physiological salt-water.

Physiological Salt-water. Dissolve 123 grains of the physiological salts in one quart (liter) of water.—It is a refreshing, agreeable beverage, preserving the health. For normalizing the serum of the blood it excels all the varieties of natural mineral water, if in addition to it daily 8 grains of amorphous silicic acid stirred into half a cup of milk or into a plate of any kind of soup, are used. This most natural of all the correctives of the blood is useful in any and every kind of acute or chronic disease. In order to supply at the same time serum for the blood and nutriment for the red corpuscles patients should use it mixed with equal parts of warm milk. On account of diphtheria these salts ought to be in every family.

Rhubarb. The root of rhubarb contains in itself a whole treasury of medicines. Its contents are as follows: About 9½ per cent of the glucoside Chrysophan, 9% of the glucoside Tannin, 11% sugar, 4% starch, 14% gum-glucoside, 4% vegetable albumen, 14% sulphate, hydrochloride, phosphate, tannate and oxalate combinations of potassa, lime and iron and 1% silicic acid. The remainder is cellular tissue and water.—To be used in disturbances of the digestion, lack of appetite, inactivity of the intestines, in costiveness as well as in diarrhoea, 4 grains of powdered rhubarb to be administered in a wafer.—I explain the action of the rhubarb-root now approved for centuries, by its rich contents of glucosides yielding oxygen, and by the absence of soda and magnesia. Only potash and lime are found in it; it seems that these two basic constituents exercise a tensive action on the albumen of the blood, which is essentially a soda-albumen, according to the law of attraction between counterparts. While oxygen is at the same time liberated from the glucosides, this quickens the nervous fibrils of the alimentary canal, and since this electric excitation extends through the walls of the blood-vessels, liver, stomach and spleen are enabled to be benefitted thereby.—In hypochondria, melancholy and other states of depression of mind, rhubarb is a sovereign remedy.

Silica, Amorphous. In connection with the use of flowers of sulphur, Calcium-magnesium-phosphate and hematite together with physiological salt-water, it is useful for asthma, epilepsy, obesity, falling out of the hair, cutaneous eruptions, nervosity, rheumatism, fluor albus and diabetes.—The effect on the nutrition of the hair is shown usually after 8 days' use by the hair's ceasing to fall out. It assists in giving a normal constitution to the connective tissue and the muscles. Every day one to three times 8 grains stirred into a cup of salted milk, coffee with cream, or soup.—It may be omitted if daily three times a dose of 15 grains of physiological earths is used in a similar manner as a corrective aliment of the blood.

Sulphur. Eight grains of flowers of sulphur are to be taken with 4 to 8 grains of hematite in a wafer in all those affections enumerated under "hematite."

Superoxide of Hydrogen. In this substance one atom of oxygen is combined in electric tension with the atoms of water. This electrizing tension is transferred by the preparation to the tissues which it touches, so that their chemical decomposition ceases. Very evident is this action in the case of a *"Cold in the Head"*. If diluted Superoxide of Hydrogen is snuffed up into the nose, the cold is frequently at once removed. The preparation has at the same time a disinfecting action; therefore *suppurating* and *putrescent* wounds when washed with it, become inodorous and clean, and heal in a very short time after laying on a dressing with oil. Whoever has once used this remedy for this purpose, will ever after know how to treasure its excellent qualities.

Ill-smelling sick-rooms become inodorous, if the concentrated preparation is used and discharged into the air with an atomizer while walking through the room; the explanation is, that the ill-smelling materials are chemically burned into inodorous carbonic acid, water and nitrogen, through the oxygen in electric tension; this is surely more rational than whitewashing with perfumes.

In every kind of *fresh* wounds, the preparation produces a rapid healing, almost while looking on.

Also internally this remedy is excellent in incipient *catarrh* of the *wind-pipe* and *bronchia,* in *intestinal catarrh* and in *catarrh* of the *bladder.* Even chronic cases of the latter ailment are cured in a day.

In *consumption* it supports the restoring effect of tonic lemonade. It suffices to use daily thrice about 231 grains of the diluted solution, which consists of 1 part of the concentrated preparation (corresponding to a 10 fold volume of oxygen) and 4 parts of water.

Also in *scarlet-fever* I have found repeated doses effectual, accompanied, indeed, with frequently repeated rubbing of the body with vinegar, as often as the redness of the skin returns.

Against a suddenly arising *diarrhoea* from taking cold often a single dose of 231 grains of the five fold dilution (1:4) proves effectual.

Nowadays the preparation which formerly easily decomposed, is mixed with small quantities of phosphoric acid, whereby without losing any of its efficacy it becomes fixed and stable and thence reliable in its application.

In using the preparation for cleansing wounds the brush or rag must not be dipped into the preparation, because it would decompose it; the preparation must be poured directly from the bottle over the brushes, the lint or the compress.

Sweet-oil (Oleum Provinciale). Every hour or every two hours one tea-spoonful, in consumption or dropsy.—The carbo-hydrates enter into combination with the urea circulating in the blood, and prevent its decomposition into carbonate of ammonia.

Tea for Consumption or Cough. Marsh-mallow root, licorice root, figs cut up, St. John's bread (Siliqua dulcis) and whole linseed, equal parts of each. These substances rich in glucosides act by giving up their oxygen to the nerves of the viscera and at the same time by giving warmth, which from the hot infusion benefits the intestinal canal. It forms a useful support in treating consumption and all catarrhal affections connected with cough.

Tea, Sudorific. A hot infusion of sheep's yarrow, chamomile, elder-leaves, linden blossoms, balm, rosemary, or wild caraway is effective in colds, partly through the warmth supplied to the internal serous membrane, partly through the nerve-quickening ethereal oil peculiar to herbs. As the circulation is thereby reduced into order, these infusions remove cramps, and are therefore especially to be recommended in uterine spasms, arising from colds or disturbed circulation.

Tonic Lemonade. Three heaped tea-spoonsful of sugar, 1 tea-spoonful of "Hensel's Tonicum" to a tumbler full of water.—Daily one or two tumblers full; besides one tumbler full of physiological salt-water, mixed with equal parts of warm milk, to be drunk at leisure. This is a specific against anaemia and chlorosis. As well as in all cases of debility.

Vinegar. Common vinegar mixed with equal parts of water

is rubbed with a flannel rag on back, breast, abdomen, arms and thighs. In all kinds of inflammation and disintegration of the blood.

Wash, Hensel's. One ounce 6 drams of powdered carbonate of ammonia are dissolved in 1½ pints of water and ½ pint of alcohol is added. The back, breast, abdomen, arms and thighs are rubbed with a rag moistened with this wash, and as soon as the rag becomes dry or warm, it is moistened again. This alkaline wash opens the pores of the skin, so that it can do its part in exhaling the carbonic acid. It is best used in the morning before rising. In cough and incipient catarrh it is of almost immediate effect.

INDEX.

late of lime arc only formed where the blood is deficient in formic acid, else this acid resolves them, 115. So if sulphates of lime and magnesia were present in the blood, easily soluble double salts would bo formed and the calculi would be prevented, 116. There is no use in patients with calculi abstaining from food containing lime, 116.

Cancer. The *fluor albus* is often the first step to uterine cancer, 129. Cancerous ulcerations are not ,to be cured by excision, 130.

Candle. The body of man is like a stearine candle which is being slowly consumed, being lit at a number of places at the same time, while stearine is continually supplied through a system of tubes to supply the waste, 33.

Carbon, Carbonic acid, Carbonates. Carbonates in the egg stand as counterparts to phosphates, 16. Since the liberation of carbonic acid is a source of energy, this explains the strengh displayed by aquatic animals; their accretion of fat also is explained by the fact that the detachment of carbonic acid from sugar produces fat, 43. Consumptives ought to be sent to the mountains to remove the carbonic acid from their blood, or at least to the higher and drier story of a house, 166.

Catarrhal affections, 128—131 consist in the throwing off of the fine membrane covering the serous layer conveying the lymphatic juice, 128. The catarrhal secretions possess an infectious character, 128. A hot drink at the beginning of a catarrh will effect a cure, 129. Hot grog is universally used as a preventive of catarrh, 131. Remedies to be used, 131.

Cells are found in several tissues but not in the blood plasma, 10. A cell is a mass of atoms which from a denser grouping of some of its elements on the periphery (according to the fundamental law of electricity) appears to have fixed limits, 11. Some cells are spherical, others are disc-shaped, cubical, polyhedric, conic, cylindrical, hooped or fibre shaped, 11. These atomic masses are held together by the law of attraction between oppositely electrified bodies and of repulsion between like electrified bodies, 11. A

cell having shape must have different sides and poles, so that dissimilar poles come together, so that connected rows are formed and around them as a central axis successive layers of similar material are formed, 11. This accumulation is limited by the zone of activity of the electric exciting centre, 11. This centre acts for some distance as a repelling power, but this is rendered complete by its counterpart: the power of attraction with regard to bodily substance, 12. The limited distance to which the power of attraction extends, accounts for the diminutiveness of the cells, 12. All space may be conceived of as filled by such atomic masses or cells, 12. The circumscribed spaces of the cells are not hermetically sealed but there is a lively intercourse between them and the outer world, 13. When the atomic mass of the cell becomes too dense it divides into two homogeneous parts or into layers of a different nature, 13. There may also result a greater number of cells, 13. A cell which is not productive loses its vital force, 14.

Centaury. The lesser centaury is useful in inertness of the uterus, 181.

Chemistry throws light on the most difficult problems of animal economy, 1. The chemical standpoint renders the foundations of the art of healing a common possession, 16. We must study the bodily structure from the chemical standpoint, 24.

Chlorine, Chloride. Hydro-chloric acid is found compounded with bases in the white of the egg but not in its yolk, 16. Chlorides stand related as counterparts to the phosphates, 16.

Chlorosis see **Anaemia,** 119—125.

Cholera. Cause of Indian Cholera on the Bay of Bengal, 84. Native cholera (nostras) due to moist autumn air, is milder, 84. Cholera naturally comes to Cairo, 85. Cholera disappears with a plenteous fall of rain, on account of the electricity thus liberated, 85. Cholera diminishes with a good overflow of the Nile, 85. Cholera is distinguished by the decomposition of the blood into lymph and fibrous tissue, causing discharges like rice-water, 86. Cholera through the coagulation of fibrin causes strange muscular contractions, 88. In cholera there is blood poisoning through

in epilepsy as shown in the case of Napoleon, 4. Earthy constituents and their salts increase electric tension and lengthen life, 76.

Egg. The egg is typical of the foundation of the mammalian organism, 16. The albumen of the egg corresponds to the albumen of the blood, the yolk to the nervous system, 16. The white is soluble in water, the yolk only in so far as the water is intermixed with solutions of salts, 16. In the white of eggs the water amounts to seven eighth of the whole, 16. Both the white and the yolk contain potash, soda, lime, magnesia and iron; in the yolk these bases are combined only with phosphoric acid and silica, but in the white with sulphuric acid, hydrochloric acid and carbonic acid (i. e. with sugar), 16. The combustible substance of the yolk is protected by a covering rich in water against the oxydizing tendency of the air, every separate fat particle is besides covered by a protecting membrane, 17. One of the most characteristic constituents of the phosphoric yolk fat is a compound of saccharine matter with stearate of glycerine and phosphate of ammonia, 17. The yolk fat may be considered as a product of secretion from the white of the egg, 17. The albumen consists essentially of equal parts of grape sugar (glucose) and gelatine sugar (glycocoll or glycin), 17. As soon as the coming together of the blood albumen and the nerve albumen produces an electric impulse, the originally homogeneous substance re-arranges itself into groups of an opposite constitution, namely ray-like filaments and a salty lymph, 24.

Electricity. Life means being electric; of such an electric state there are innumerable degrees, 14. The measure of electricity given us may be augmented or diminished by external means, how? 15. We must regard the cortical stratum of the brain albumen as radiating electricity, like a conductor giving sparks, 24. Every species of motion can be traced back to electricity, and even our vital movements are derived from this common source, 40. Chemical changes also are due to electricity, 41. Solar electricity causes albumen to be formed, consequently the de-composition of albumen must again set the solar force at liberty, 41. Health is the continuance of the cohesive electric force, illness its diminution, 62. Electrical energy is mainly produced through the combination of the air respired with the ferruginous gelatine of the blood, 65. Electricity proceeds along the spiral filaments of the sympathetic nerve, which help to form the wall of the arteries, right on to the capillaries and magnetizes the blood globules flowing through, 65. Three things contribute to electricity with man, the oxygen respired, the saltiness of the blood, and its earthy constituents, 72. Continuous radiation of the electricity in the nerves causes convulsive twitches as in St. Vitus's dance, 102. Diminution of the electric fluid is the real cause of disease, 105.

Elephantiasis 150—153 is caused by gross meat-diet and much wine drinking, 150. It is due to the elimination of sulphates from the body and the predominance of phosphorus, 150.

Emotions occupy the first place among factors causing disease; their action amounts to an electro-chemical process, 6.

Epilepsy caused by lack of earthy constituents in food, 4. Epilepsy is often owing to insufficient nutrition, 101. Convulsive twitchings are caused by the electricity stretching the nerves and the surrounding muscles into straight lines, 102. A case caused by an injury to the pneumo-gastric nerve, 103. Epilepsy caused by interrupted catamenia, 103. The crawling sensation preceding an epileptic attack is owing to a withdrawal of the electric induction current, 103. Disturbances in the circulation give rise to epilepsy, 104. *Epilepsy* 139—143.

Equivalence of Force demands the chemical dissolution of every organic structure at the same time that it produces warmth or motion, 25. Equivalence of force shows that there is really but one form of force, which I would call "polarity", but as this term is not familiar we will call it "electricity", 40.

Eruption, cutaneous, 150—153.

Faraday built an electric chamber, 137.

Fat. Nearly one half of the yolk consists of fat, *ergo* of combustible substances. 17. Fat may be produced from sugar and starch and from albumen, 56. Unhealthy accumulation of fat. 144.

Fatigue is felt when the presence of carbonic acid interferes with the oxidation of nerve-fat, 46.

Fermentation takes place on the removal of carbonic acid and electricity from sugar; in this decomposition the electricity not employed in building up a new body assumes an altered form and manifests itself as heat, 42.

Fever. Fever heat resembles the warmth produced by fermentation; thence in fever a much greater amount of urea is found in the secretions, showing an intense decomposition of albumen, 46. *Fever* 144—149.

Fibrin. Fibrin, albumen and lymph originating from the albuminous matter give us the key to the structural organization of our body, 24. Nothing is needed for the formation of fibrin from albumen, but a blow or impact and the admission of oxygen, 28. Fibrin mainly contains gelatine sugar or glycocoll, 28. The tendinous fibre draws further material out of the albumen, according as it is subject to tension, 29. The fibres are similarly hardened in old age, 30. The baby spins fibre or silk out of the albumen every time it stretches its feet or waves his little arms, 30.

Foehn causes in advance headache and sleeplessness, 84.

Formic acid is found in all healthy blood and dissolves the phosphate and oxalate of lime which else form calculi, 115; useful in consumption, 165. Combines with the urea and prevents its transmutation into carbonate of ammonia 183.

Fungus grows out wood when the easily soluble sulphates have been dissolved out of it, 151.

Gall-stone colic, 166.

Galen in teaching commenced with the skeleton because his father was an archi-

Hensel, Macrobiotic.

tect, we have not advanced as yet beyond Galen, 8.

Gall-fly. The larva of the gall-fly is doubtless produced by the meeting of albumen in the leaves of plants with an atom of phosphate of lime brought by the sap, 21.

Gelatine. The gelatine of the cartilage must have been formed from dehydrated saccharine matter and ammonia, 28. Gelatine sugar owing to its containing glycocollic acid and ammonia can combine both with acids and with bases; thence it is that salts can be assimilated by our organism, 29. The characteristic chemical basis of the body is gelatine which glues bones, muscles and nerves together, 63. This basis is formed from gluten, milk, albumen and from the gelatine of bones, and is the source of the vibrating elasticity of the body, 64. The net-work of gelatinous matter extends throughout the whole body and this gelatine may putrefy even during life as in small-pox etc. 65.

Generation. The *original* generation of animal life at least with respect to articulated animals is solved by the theory of the origin of tannin and of lecitin from saccharine matter, 21. Thus may be explained the origin of gall-flies, wood-lice, phylloxera, caterpillars and the larvae of bark-scarabs, 21. 22. If sunshine is lacking, only mites appear in the moist flour; if but little ammonium phosphate is present, various kinds of lice, such as plant-lice, root-lice, phylloxera are produced; if air is lacking, we get the marine phosphorescent sea-nettles; with salt and limestone there are formed crustaceans and oysters, 22.

Gentian animates the stomach, 181.

Gland. A special gland determines the decomposition products into which the blood shall be converted, 69.

Glauber's salts with vinegar makes impossible the spread of most of the acute diseases, 105.

Glucosides are substances that can give off oxygen when taken into the blood, 160. Among these are Gum arabic, gluten, licorice, 161; also marsh-mallows root and linseed, 162. Mixtures of these glucosides are also prepared, 162. 163. Among

13

the lungs must be prevented by supplying more oxygen, 159.

Lecitin or nerve fat is composed of stearate of glycerine and ammonium phosphate, 17. The compound of Lecitin and sugar is called Protagon, 17. Lecitin may under certain conditions be formed from saccharine matter, 21, Lecitin is an essential basis of the more intelligent animal life, 22. The greater acuteness of the senses of bees, ants, flies etc. is due to the greater purity of their lecitin, 22. Pure lecitin requires for its production 7 molecules of sugar, 1 of ammonium phosphate, 8 of oxygen, 42 of water, and sunshine, 22. Lecitin in combination with saccharine matter is the basis of innumerable different kinds of albumen, as sugar combines variously with earths and salts, 22.

Lemonade useful in dropsy, 171.

Leprosy, 150—153 is owing to improper nutrition, as seen in India, Hawaii, Norway, Italy and northern Germany, 151. Lepers are recruited from those whose food is deficient in sulphur, lime, iron, silica etc., 152. Leprosy is an intensive scrofula, 152.

Lice. Various kinds are produced in the conversion of saccharine matter into lecitin, when little ammonium phosphate is present, 22.

Lichens and mosses which show no sulphur in their ashes grow out from the bark of trees, 151.

Licorice furnishes free oxygen to the blood, 161. *Licorice-drink*, 184.

Life. The life of our body consists in chemical combustion, 1. Life means being electric, productive; it means acquiring, forming, building up and calling into existence, 14. Lime, either alone or combined with sulphuric acid and the halogen acids, can enter into chemical condensation with sugar, similarly potash, soda, magnesia, manganese and iron, thus giving rise to a variety of vegetable and animal forms of life, 22.

Lime and sulphur are indispensible to man but injurious to worms, 5. Without lime firm bones cannot be formed, 73. Lime represents a certain quantity of electric cohesive force, 73. Lack of lime converts the intestinal epithelium of chil-

dren into worms, 74. Precipitated Carbonate of Lime in cholera morbus, 184.

Linseed is useful in furnishing free oxygen to the blood, 162. •

Lymph, Lymphatics. Fibrin, albumen and lymph, these three substances originating from the albuminous matter give us the key to the structural organization of our body, 24. The threefold division into nerve-substance, fluid lymph and hard tissues is of universal application to our whole bodily organism, 24. The lymphatics have been hitherto regarded largely as carrying merely waste material; on the contrary the lymph is the noblest material in our organism, 30. The lymph contains little gelatine sugar, but besides albumen it contains the sugar material and the earths and salts that are not used for the production of fibrous tissue; its richness in sugar enables it to supply the materials for the formation of fat, which takes place by separating carbonic acid from the sugar, 31. Through its containing ammoniacal albumen, phosphatic earths and salts, saccharine matter and fats, the lymph nourishes both blood and nerves, which thence supply their waste, 31. The lymphatics and the lymphatic glands supply fresh material to the blood and the nervous system, which shows the perniciousness of the practice of removing swollen lymphatic glands with the knife, 31. The fluids in the lymphatics require 6—8 hours to replace the waste in the nerves and blood, 47.

Madagascar has yellow fever, why? 84.

Maggots are produced in ham-bones, by exchanging the lime in the gelatine of the bone for ammonia, thus changing the bone gelatine into living nerve-material, 32.

Magnesia accompanies lime almost everywhere, also in the bones, 110.

Marchpane is of great value in furnishing oxygen to the blood, 163.

Marsh-mallow root furnishes free oxygen to the blood, 162.

Marsh-trefoil useful in liver troubles, 181.

Marshy districts are poor in oxygen, and deficient in electricity, causing deficiency of nerve action, 83.

13*

20 oz. of oxygen or more than 5000 quarts of air are required for one day's work in a temperate climate, 78. The diminution of oxygen in yellow fever, causes prostration, black vomit and confusion of the brain, 82. The glucosides at the same time furnish oxygen to the blood and fat, 160.

Paralysis, Paralyze. The inadequate amount of oxygen in the blood causes a gradual paralysis of the nerve fibrils, 81.

Perpetuum Mobile(a) in the body is the Spleen, 59.

Perspiration when abundant impedes the secretion of urine and as the urea is then retained in the blood, malaria may result, 83. 84.

Philadelphia. Cause of yellow fever there, 84.

Phosphorus, Phosphates. Phosphates stand as counterparts to the carbonates, chlorides and sulphates, 16. Phosphates are found in the yolk but not in the white of the egg, 16. Saccharine matter compounded with phosphate of ammonia and stearate of glycerine in the fat of the yolk, 17. Phosphate of potash is almost the only mineral base in meat, wine and beer, 114.

Phylloxera originate by spontaneous generation in root fibers by the conversion of saccharine matter into lecitin, 22. Phylloxera are produced when there is little ammonium phosphate present, 22.

Physiological Earths, their constituents, 174. Needed to supply new hemoglobin when destroyed by excessive use of wine, 174. where to be procured 174. see also 184.

Physiological Salts are formed in faithful imitation of the salty contents of the blood, 106. By drinking the physiological salt-water the most natural means of transfusing blood is provided, 106. Where these salts may be procured, 106 note. Used to cure sores, 124. Their use in fever, 146; in consumption, 164. Their constituents, 184.

Physiological Salt-water, 184.

Physiology teaches that our bodies are made up of C, O, H, N, Cl, F, P and S; all substances which can be converted into gases, the earthy materials whose function is unknown being largely ignored, 3.

Piles are frequently associated with asthma and both are due to insufficient oxygen in the blood, 109. How piles may be cured, 109. Treatment, 150.

Plants. The shapes of plants alter when deprived of substances they need, 23.

Pneumogastric Nerve produces so close a relation between the brain and the organs to which it extends, that we see vexation cause enlargement of the liver, want of appetite follows on trouble, grief and care, micturition on anguish, fright and terror etc. 44. 56. An injury to this nerve causing epilepsy, 103.

Poppy-seed useful in consumption, 162.

Potatoes are rich in earthy materials, their introduction into Germany has therefore put an end to plague and leprosy, 152. The constituents of the ashes of potatoes given, 152.

Prostration ensues, when the urea remains in the blood and by taking up water is changed into ammonium carbonate which paralyzes the nerves, 81.

Protagon is a compound of Lecitin and sugar, 17. 23.

Prussic acid may be produced in the body by withdrawal of water from the partially oxidized hydrocarbons of gelatine and ammonia, 67. Prussic acid compounds may by the action of water and acids or basis be reconverted into harmoless fatty combinations, 68. Prussic acid turns the iron of the blood into blue-black prussian-blue in yellow fever, both in the stomach and the intestines, 81. Prussic acid is internally generated in cholera in great quantities, causing sudden death, 88.

Ptomaine is the poison of putresence and is found in the sputa of consumptives, 99. This ptomaine spreads the putrescent decomposition to neighboring tissues, 100.

Pulse, Threadlike pulse in cholera, 87.

Quassia is brief in its action, but gives tension to the intestinal nerve, 182.

Racahout is highly useful in consumption, 164.